Integrating a Usable Security Protocol into User Authentication Services Design Process

T0225353

Integrating a Usable Security Protocol into User Authentication Services Design Process

By
Christina Braz
Ahmed Seffah
Bilal Naqvi

CRC Press
Taylor & Francis Group
Boca Raton London New York

CRC Press is an imprint of the
Taylor & Francis Group, an **informa** business
AN AUERBACH BOOK

CRC Press
Taylor & Francis Group
6000 Broken Sound Parkway NW, Suite 300
Boca Raton, FL 33487-2742

First issued in paperback 2020

© 2019 by Taylor & Francis Group, LLC
CRC Press is an imprint of Taylor & Francis Group, an Informa business

No claim to original U.S. Government works

ISBN-13: 978-1-138-57768-8 (hbk)
ISBN-13: 978-0-367-65692-8 (pbk)

Visit the Taylor & Francis Web site at
http://www.taylorandfrancis.com

and the CRC Press Web site at

http://www.crcpress.com

Contents

Why this Book?

Most often, security technology and service development don't do user research. We often hear people say things like: "The most important issue is the powerfulness of the technology to secure systems and information." Or: "Usability research limits or is wasting time." And: "It's not even necessary because the developers are themselves part of the community of users and thus instinctively empathetic to what those other users find useful or usable."

You, clearly, think otherwise. You think it's important to know who is using the products you're making. And, you know, you're right. Finding out who your customers are, what they want, and what they need is the start of figuring out how to give it to them. Your customers are not you. They don't look like you, they don't think like you, they don't do the things that you do, and they don't share your expectations, assumptions, and aspirations. If they did, they wouldn't be your customers; they'd be your competitors.

This book is designed to help you bridge the gap between what you think you know about your users and who they really are. It's not an academic treatise. It's a toolbox of concepts to understand how people experience products and services. The techniques – taken from the worlds of human–computer interaction, marketing, and many of the social sciences – help you know who your users are, to walk in their shoes for a bit.

In addition, the book is about the *business* of creating usable products. It acknowledges that product development exists within the complexities of a business venture, where the push and pull of real-world constraints do not always allow for an ideal solution. User research is a dirty business, full of complexities, uncertainties, and politics. This book will, if it serves its purpose, help you take some of that chaos. It will help you gain some clarity and insight into how to make the world a little better by making products and services more thoughtfully.

Who Are You?

This book was written for people who are responsible, in some way, for their products' user experience. In today's digital product and service development world, this could be any number of people in the trenches. In fact, the responsibility may shift from person to person as a project progresses. Basically, if you've ever found yourself in a position where you are answering for how the end users are going to see the thing you're making, or how they're going to interact with it – or even what they're supposed to do with it – this book is for you.

This means that you could be:

- A program manager who wants to know how to prioritize a team's efforts
- A designer who needs to create and refine new ways to interact with and through digital information
- A marketing manager who wants to know what people find most valuable in your products
- An information architect who needs to pick an organizational scheme
- A programmer creating a user interface, trying to interpret an ambiguous spec
- A consultant trying to make your clients' products better
- An inventor who wants to make a product people will use

Regardless of your title, you're someone who wants to know how the people who use the product you're making perceive it, what they expect from it, what they need from it, and whether they can use what you've made for them.

What's in This Book?

This book is divided into three major sections. The first section (Chapters 1 through 3) describes why end user research is good, how business tensions tug at the user experience, and presents the state-of-the-art of user authentication methods.

The second section (Chapters 4 through 6) presents a philosophy for creating useful, desirable, usable, and successful products. It also contains a short chapter on a technique that will teach you in 15 minutes everything you need to know to start doing usability research tomorrow. Really. It is also a cookbook with a dozen techniques for understanding people's needs, desires, and abilities. Some of the chapters are completely self-contained, such as the chapters on surveys and usability tests. Others describe supplementary activities, such as collage and map making, to use in conjunction with other techniques. We don't expect you to read these chapters in one sitting, in order. Far from it! We assume that you will pick up the book when you need it, reading chapters to answer specific questions.

The third section (Chapters 7 and 8) describes how to take your results and use them to change how your company works. It gives you ideas about how to sell your company. Best practices in research change quickly, as do preferred tools. We have moved much of the reference material in the previous edition to the book's website.

What's Not in This Book?

This book is, first and foremost, about defining problems. All the techniques are geared toward getting a better understanding of people and their problems. It's not about how to solve those problems. Sure, sometimes a good problem definition makes the solution obvious, but that's not the primary goal of this text.

We strongly believe that there are no hard and fast rules about what is right and what is wrong when designing experiences. Every product exists within a different context that defines what is "right" for it. A toy for preschoolers has a different set of constraints than a stock portfolio management application. Attempting to apply the same rules to both of them is absurd. That is why there are no guides for how to solve the problems that these techniques help you to define. There are no "top-10" lists, there are no "laws," and there are no universally reliable heuristics. Many excellent books have good ideas about how to solve interaction problems and astute compilations of solutions that are right much of the time, but this book isn't one of them.

Acknowledgments

We'd like to thank the companies who provided material, some previously unpublished, for case studies: Adaptive Path, Food on the Table, Get Satisfaction, Gotomedia, Lextant, MENA, Design Research, PayPal, Portigal Consulting, User Insight, and Users Know. We would especially like to thank our reviewers: Todd Harple, Cyd Harrell, Tikva Morowati, and Wendy Owen. We'd also like to thank the people who have generously given us advice and help, including Elizabeth Churchill and Steve Portigal.

And, of course, our families, who put up with us throughout the very long writing and revision process.

Chapter 1

Usability and Security: Conflicts and Interdependencies

1.1 Introduction

In many software products, systems, and services, human users are a critical part of the security process; for example, they create and use passwords, follow or have to follow security protocols, and share data that can impact a system's security, both positively and negatively. However, most often, many security concerns are designed with little or no attention paid to the usability concerns or to the human user's cognitive abilities, user experiences, workflow, and tasks. As a result, people find ways around the security obstacles that get in the way of their work.

It is increasingly being identified from both academic research as well as industry practices that usability and security are intimately linked quality attributes. The linkage between the two is mutually antagonistic or, in other words, conflicting. Despite the increasing awareness and identification of conflicts between the two attributes, the state-of-the-art is not well aligned due to many factors.

Firstly, the lack of commitment to usability in the early design stage has been reported by several studies. As a matter of fact, Theofanos (2006) reported two examples of software project failure due to usability problems. One example is a $46 million computer system by the US General Services Administration (GSA) regional offices in Denver and Philadelphia in 2004. The new system that was designed to improve financial management was unnecessarily complicated to use. For example, due to security concerns, instead of being able to save a file with a few clicks, employees were required to learn 15 steps. The second example is a similar experience in a UK passport office. After installing a new system for issuing passports, a backlog of passports started building up and it led to a delay of up to three months to obtain a passport. The authors report that reasons for loss of productivity included the large number of keystrokes and onscreen operations required. Constantine and Lockwood (1999), in their seminal book *Software for Use*, also provided several examples that highlighted a major problem with current usability methods.

Secondly, software and user interface designs are developed in a haphazard way, based on tens to hundreds of imprecise and conflicting usability guidelines, heuristics, and thousands of design "tips or guesses", with few of them drawn from empirical evidence or formal proofs. Usability evaluation is not only performed late in the design process, but it is also expensive, time-consuming, and the results are sometimes rough and qualitative. Thus, the refinements made to the original design are highly controversial; in addition, they make designers reluctant to explore innovative but risky alternative design solutions. What is needed is a model that predicts how users perceive the design usability in real life, thereby giving designers immediate feedback on their early design concepts and the ability to compare between different design alternatives.

Thirdly, it has been reported that software technology alone will not provide all of the solutions to security problems. Human factors (usability first and foremost) play an important role in keeping systems secure, and it is important for security and privacy experts to have an understanding of how people will interact with, use, and abuse the systems they develop.

The questions that rise in this concern are, can we – software security developers, usability, and human experiences designers – predict usability of security features from the early design artifacts and models such as low-fidelity user interface prototypes, specifications of software architecture, use cases, etc.? Can we define objective predictive measures to evaluate usability in the early software development stages? Can the measures of usability prediction be correlated with the results of evaluations made by experts using criteria that are most of the time inconsistent, difficult to understand, by developers who have not been trained in human–computer interaction (HCI)?

The practical context of this book can be seen in the future Internet – meaning the Internet of information and services, the Internet of things, the Internet of people and online communities, as well as the underlying cloud computing infrastructure. The Internet today is becoming the integrated infrastructure for supporting people in their everyday life and work, as well as companies in their everyday business. Within this proposal, cloud computing is defined and understood as a paradigm shift from traditional installable applications to web-based software, where applications live on the Web as services. They consist of data, code, and other resources that can be located anywhere in the world, on remote servers "in the cloud". Cloud computing is a step toward the future vision that computing will be a utility similar to water, electricity or telephony (Buyya et al., 2009). We have adopted the National Institute of Standards and Technology definition of cloud and Web services (NIST, 2011):

> *The capability provided to the consumers, users, and stakeholders to use the provider's applications running on a cloud infrastructure. The applications are accessible from various client devices through a thin client interface such as a web browser (e.g. web-based email). The consumer does not manage or control the underlying cloud infrastructure including network, servers, operating systems, storage, or even individual application capabilities, with the possible exception of limited user-specific application configuration settings.*

In the context of the future Internet, companies are spending millions of dollars on security technology such as firewalls, encryption, and secure access devices, but most of the time they forget to address issues related to the weakest link in security engineering: the human experience and the usability concerns. The front end of the service showing to the user should be designed so that it is suitable to the risk involved and as easy to use as possible. Applying too low a level of security might compromise the integrity of the company's process. But applying too high a level for a low-risk process means the process will be too hard to use and will confront low usability rates. As stated by Nagel et al. from Forrester (2008), the key criteria when assessing secure system features are ease of use, portability, cost, security, manageability, and cross-channel utility.

Figure 1.1 portrays the intimate cause–effect relationships between usability and security. One typical situation that illustrates this intimate relationship is the user authentication service, which is one of the basic services incorporated in

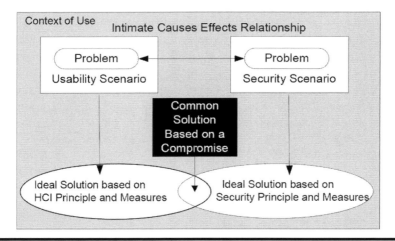

Figure 1.1   The Usability and Security required trade-off.

today's Web and cloud platforms (citizen online services, e-commerce websites, online community tools, etc.). Indeed, security and usability are both essential in user authentication and the underlying identity management services. One of the biggest challenges facing heterogeneous organizations is providing usable and secure access, authentication ("who do you claim to be"), and authorization ("we will grant you these rights") of users to systems. Another particular concern in authentication according to Cranor and Garfinkel (2005) is that authentication technologies do not fail gracefully. Failing gracefully means that even if authentication fails (e.g. user forgets his/her username but gets the password right) the system can give him/her partial access to the service or secure and fast support in getting login data to a safe place.

This common human-centric vision of the future Internet, which needs much more than just reliable and secure back end applications and communication protocol; it requires that the designers will have to develop and adapt the front end services to the experiences, capabilities, behaviors and usability needs of the stakeholders and end-users of the services. The belief that security and usability are two opposed quality factors has to be avoided. Even if security and usability are related to different components of a system or a service (UI and the functionality) (Cranor and Garfinkel, 2005; Jøsang et al., 2007; Nielsen, 2000) and even if the security and usability can be engineered by two separate teams, mainly software engineering and human–computer interaction teams, there are several cases in which security and usability should be enhanced by modeling their intimate mutual relationships. Typical examples include online payment and e-banking services, supervision systems of critical industrial infrastructures, crisis management, and rescue services. More attention should be paid to the front end of these solutions that have to be very secure, i.e. how they show to the user directly and indirectly. Usability cannot be treated separately from the security engineering of the whole system.

Our perspective is also grounded in the field of human-centered software engineering that looks at avenues for closing the gaps between HCI and software engineering (Seffah, 2006; Seffah et al., 2008, 2009). User interface developers and service engineers should have the same knowledge in HCI and user-centric design, and tools for implementing designs correctly. This knowledge will result in usable front end design, reduction of the number of security bugs, and lower development cost per service. Both user interface designers and service developers should be assigned responsibility for the accurate implementation of front end design, as well as back end service functionality. We think that service developers should understand user interface issues sufficiently well to know when to raise design issues during implementation, rather than disregarding them or implementing them inaccurately.

The problem is not just about the usability of the user interfaces of security tools, but security requirements, user experiences design and, most importantly, user involvement in security design and engineering. According to Whitten and

Tygar (1999), most of the research focuses are on providing better UIs, but it is obvious that usability problems with secure systems are more than only UIs and the need of HCI factors and design methodology. The authors claim that using conventional methods for usability evaluation only assess the usability impact on security effectiveness. Both analytical and empirical evaluation was performed in testing the usability goals of Pretty Good Privacy (PGP) (Whitten and Tygar, 1998), a public key encryption program. A number of usability problems causing security failures were discovered in the study, providing the foundation in case that specific usability goals are needed for usability evaluation of security mechanisms (Whitten and Tygar, 1999).

1.1.1 The Interplay between Usability and Security as Key Quality Factors

Security and usability are quality characteristics that affect the quality of software products. The term *usability* refers to multiple concepts such as execution time, performance, user satisfaction, comprehensibility, security, ease of learning, understandability and so on. There are several existing standards related to usability, and they can be classified into four different categories (Abran, Khelifi, Seffah et al., 2003): Product effect (output, efficiency, and satisfaction in the moment of using the product), product attributes (interface and interaction), the process used to develop the product and organizational capacity.

The benefits of usability have been also explicitly demonstrated for the Internet (Bias and Mayhiew, 2002). These include improving productivity, reducing the costs of the training period and to develop documentation, increasing user satisfaction, etc. According to Donahue (2001), the cost–benefit ratio for usability is valuable because every dollar invested in usability gets $30.25. Over the years, the HCI community has developed many design techniques to ensure usability. However, these techniques have mostly been developed separately from software engineering community, which makes it difficult to utilize the approaches. Seffah and Metzker (2004) mentioned that there are five major obstacles from the perspective of both the software engineering community and from the community of usability engineering:

1. User-centered (UCD) and usability engineering techniques are not connected with the lifecycle of software development.
2. There is an important cultural gap between the practices of software engineering and usability engineering.
3. Software developers do not use usability engineering tools but develop their own, resulting in many cases in the reinvention of existing tools.
4. It is necessary to train professionals in usability and software to work together.
5. We need a framework to promote and improve the techniques of UCD and software engineering proposals in the two communities.

Computer security has been a field that has grown tremendously since the '70s, leading to a lot of techniques, models, protocols, etc., which have been also accompanied by a pronounced activity of international standardization and certification organizations. In fact, as stated by the International Telecommunication Union (ITU), there are many international standardization organizations that have produced a complex structure of standards relating to computer security issues, which change and are updated very frequently. This makes security's definition somewhat confusing.

One traditional definition is from Castano et al., (1995), which defines it as "protection of information processed by a computer against unauthorized access, modifications, improper or lack of availability of a service in a given time". Another classical definition is that offered by the International Organization for Standardization (ISO)9241–210:2010, which considers security as a sub-factor of software quality and defines it as "the ability of software products to protect data and information against unauthorized access or modifications, ensuring the access is not denied to authorized users". More recently, the ISO 27000 series of standards have been specifically reserved by ISO for information security matters (ISO/IEC 27001, 2013). Security has been defined as the inability of its environment to have an undesired effect on the system. The preservation of confidentiality, integrity, and availability of information, in addition to other properties such as authenticity, accountability, non-repudiation, and reliability, can also be involved.

While security has been interpreted as a purely technical aspect in software development methodologies, some authors think it is much more than that, taking instead a strategic dimension, resulting in one of the most important criteria in the governance of ICT (Posthumus and Solms, 2004). Additionally, there are many security-related concepts (models, techniques, policies, services, requirements, etc.), which often intermingle and increase the complexity of their understanding, and in many cases the same term may be applied many different ways.

All these advances in terms of perceptions and models of security and usability are very important, but they are most often completely independent from each other. That is why it is necessary to make an effort to develop a unifying model that describes the intimate relationship between usability and security.

1.1.2 Background

1.1.2.1 How Security Engineers Addressed Usability/HCI Concerns?

One reason that explains the failure of security specialists to address usability as perceived and defined by the HCI community issues is that security and usability have historically evolved independently or have been considered as two opposites factors. Another historical explanation is that researchers were driven more by technology rather than by user problems and perceptions of security. For example, the development of identity management technologies was so demanding in terms of security that it left little time and money for usability and the human factors in general.

A second factor that may be advocated is the industry's behavior in being more driven by bug fixing rather than trying to examine and consider the context and the user experiences in which the bugs occur. Therefore, most industry efforts have been on automating the process of reporting and handling bugs rather than looking for human experiences and how they can promote more secure operations overall.

Another reason that demonstrates the lack of alignment between security and usability is the design and innovation approach leading to new security technologies. Most often, the innovation is initiated by a company developing an "in-house technology" that addresses a specific problem which occurs in a specific project. Other groups in the same company or in other companies may develop their own versions of these solutions. This makes it difficult to ensure the usability of these in-house solutions and several versions of them when changing the original context of their applicability. Firewalls, junk mail filters, spyware, and antivirus are good examples.

Finally, the lack of HCI skills required for conducting effective user studies is a serious obstacle. This is because it makes the results of the studies done by the academic research security community highly questionable. In industry, user studies are highly difficult to conduct because they have to take into account the regulations/laws that govern the use of human subjects in experiments related to the safety and security of systems and services.

1.1.2.2 The Lack of Deep Usability/HCI Studies on Security in the HCI Community

The term usability has been defined in different ways in the literature, which makes it a confusing concept. To illustrate this point, some broad definitions of usability from two different standards are listed next:

- The capability of the software product to be understood, learned, used, and also be attractive to the user, when used under specified conditions (ISO/IEC 9126-1, 2001).
- The extent to which a product can be used by specified users to achieve specified goals with effectiveness, efficiency, and satisfaction in a specified context of use (ISO 9241-11, 1998).

There are also varying definitions across different sets of standards or authors concerning more specific attributes (facets, aspects, factors) of usability. Recently, the usability question has been linked with security in numerous sources. The first book (Cranor and Garfinkel, 2005) combining these two issues was published in 2005. The first Symposium on Usable Privacy and Security (SOUPS) was held the same year, and since then security and usability have been seen as a suggested topic for papers in many security- and HCI-related conferences. However, the main research in this area has been done on the usability of certain security-related applications, the first example of those being Whitten and Tygar (1999).

1.1.3 Objectives and Practical Outcomes of Usable Security Research

The domain considering the human aspects related to security and the integration of usability into security features is referred to as usable security (Garfinkel and Lipford, 2014). The overall objective of usable security research can be stated as follows:

> *Identify and model the intimate relationships between usability and security characteristics in Web and cloud services, and develop concepts, metrics, patterns, methods, and tools all embedded into an integrative human-centric design framework to supporting rich user experience and usability without compromising the security of the overall services system.*

Our position is that there is a need to consider this intrinsic conflict between creating Web and cloud services that are usable and designing underlying systems and cloud computing platforms that are secure. The main focus needs to be on early design phases to make the security and usability interplay an outcome of the requirements definition and concept design phase.

This overall goal can be depicted in the following specific objectives addressed by the different work packages:

- Setting concrete targets for the user/stakeholder experiences with services.
- Detailing and modeling the related user activities and tasks, services, and the usability/security symmetry.
- Discovering and documenting design solutions and patterns that mitigate the usability security problem identified, a proven solution, and the different user–service interactions in which the problem occurs.
- Providing metrics and tools to assess objectively the level of security and usability as quality factors.
- Integrating the four previously-listed objectives into an integrative human-centric framework.
- Instantiating this framework to different case studies.
- Ensuring the standardization and the long-term sustainability in industry of the integrative framework as well as the avenues of its integration into industry design methods and tools.

The practical measurable consequences include:

- *Finding the right trade-off between security and usability as early as possible in the design and engineering lifecycle.* Usability problems in secure Web systems can lead to security vulnerabilities which can consequently impact a *company's bottom line.* One of the difficulties in developing human interfaces to security systems is anticipating the response of users to the huge space of possible system states and design options. The representation of user activities and tasks with related user experience targets will allow the designer to simulate user responses to a diverse range of situations and design options.
- *Establishing a solid theoretical ground for characterizing the usability security symmetry.* This should start with a robust and well-defined specification of security usability symmetry, proven solutions, task and service models; it should then move on to their usage as driving artifacts from which the implementation of the service is generated, the way it is deployed and tested.
- *Providing proven solutions as practical and standardized patterns and measures for the design and evaluation of usable secure services.* Even if the HCI and security engineering research communities have been gradually developing a good body of work in usable security, most of them are general guidelines which are not easy to apply for specific problems. Contrary to general usability design guidelines, which are mostly descriptive, and simply specify "nice to have" general design features, there is a need to develop design patterns enriched with measures, user experiences, and task models that together will provide proven solutions on how a problem can be solved.
- *Developing robust design tools and methods.* The security and usability requirements and design phase is an important prelude to extracting and gathering the user requirements. It is especially important because it defines the problem that the stakeholder is trying to solve, no matter what model of software development process is adopted (e.g. waterfall, iterative, agile, model-driven, service-oriented, etc.). It is broadly held that gathering and agreeing on requirements and design is crucial in the whole *development* and also important to any successful project. The software quality community should aim at developing robust design tools that influence the whole development process and bring security and usability together earlier in this process.
- *Integrating the measures, patterns, models, tools, and methods into a human-centric design and engineering framework as well as an open platform for supporting the standardization, integration and sustainability.* The developed tools should be available on an open platform for collaboration and open source dissemination. The standardization action and the open source platform are two ingredients for ensuring the sustainability and evolution.

1.1.4 Assumptions and Hypotheses Concerning Usable Security Research

- *Usable security is critical to the effective adoption and deployment of Web and cloud computing services.* As a matter of fact, there is no set of recognized usable security standards particularly targeted to the interplay between user authentication and usability but only to security in general. As expected, there are numerous examples that fully characterize this hypothesis, such as the so-called password complexity, locking personal identification number (PIN) systems, cumbersome data input of challenge–response calculators, lack of usability in security software, "negative redundancy" (this term has been coined by this research) of biometrics systems when users are being authenticated by a system (e.g. combine a username/PIN with fingerprint), and so on.
- *The development of secure usable services should be considered at an early stage as part of requirements and design phases.* This means it should be considered at even an early stage as part of user requirements (i.e. security services) in a trusted security framework of an organization.
- *Security designers should address the diverse needs and experiences of different human actors, including service end-users, service providers, brokers, security administrators, employees, business partners, and customers in general.* For example, security administrators might enforce the use of a strong public key infrastructure (PKI) for a certain service. Employees who want to mix physical-access security with strong network authentication might require a smart card aggregating use of a one-time password (OTP), PKI, and building-access credentials. External users such as remote virtual private network (VPN) users, business partners, and customers may prefer clientless OTP tokens. Users who conduct high-value transactions and need non-repudiation capabilities may require a hybrid authentication platform that can combine OTP and PKI functionality such as hybrid tokens from RSA Security* (i.e. all in-one-devices). Lastly, organizations selling large-scale consumer applications may prefer consumer scratch cards over more expensive electronic devices for authentication. The issues here are how to ensure the usability of all these security mechanisms as a whole and how to make sure the usability of a specific mechanism will not compromise the overall security of the service.

A wide variety of security tools have been developed, but their successful use in real applications is fairly limited because of their complexity, their "difficulty of use", and the necessity of previous advanced technical knowledge on the part of end-users. The "hard-to-use" restrains not just novice users but also the average corporate and consumer computer users. This originates from the fact that several steps and parameters have to be set on those security applications so that security services, which

---

* https://community.rsa.com/docs/DOC-77347.

range from security policy development (i.e. authentication included) to intrusion detection support, can be properly executed. However, such configuration complexity leads to unwelcome circumstances in which some users are ready to give up security to meet their project deadlines or to attain higher system performance.

Today, there is a clear understanding of the user's role in the security of any system, as just one of many links of a chain which can be considered to surround and secure the system. In the context of Web authentication, Renaud (2003) noticed that one way to make the user link stronger is to consider essential factors such as the user's needs, abilities, inclinations, and skills in formulating security mechanisms and policies. Therefore, security architects and designers need support in assessing the options to meet their specific security versus usability requirements. The development of methods and techniques to diminish complexity in the usage of security services is therefore required.

1.1.5 Progress beyond the State-of-the-Art

We identified four crucial areas that need to be considered in designing and engineering secure usable Web and cloud computing services. These areas are: (1) task and service model-driven engineering, (2) user experience/user-driven design, (3) metrics-based usability and security evaluation, and (4) design patterns and pattern-oriented design.

1.1.5.1 Service-Oriented and Model-Driven Engineering

In the area of service-based interactive applications, some work focused on the issue of how to automatically generate UIs to access Web services (Kassoff et al., 2003; Spillner, 2008). However, such approaches did not address the problem of service composition or the security aspects. Business Process Model Notation (BPMN) models are an example in which the composition of services is done only at the service level. Such models can be translated into Web Service Business Process Execution Language (WS-BPEL) (BPEL, 2007), an XML language for describing and executing business processes, which can be used for the composition of Web services at the business level. However, it does not include information on the user interface for service access.

In the research direction covering the issue of building interactive applications by using/composing reusable components, other contributions also address aspects related to multi-device support. Indeed, a multilayer architecture (Pinna-Dery, 2003) (one abstract UI and multiple concrete UIs) was proposed for promoting the reuse of UI components (e.g. merging different UI elements), and also adapting them to multiple devices. Another work dealing with the creation of adaptable user interface using a model composition approach is presented in Raphael et al. (2002). With this methodology, user interfaces are built by exploiting the composition of model fragments, stored in a library and appropriately selected in order to fulfil their needs. Each one represents a specific aspect of the graphical user interface

(GUI). The users specify the view and the control part of each fragment, and they are also able to adapt the UI. For instance, new fragments can be included, and the visual rendering of fragment attributes can be modified. However, the proposed approach only deals with graphical user interfaces.

Service-oriented architecture (SOA) and model-based approaches for HCI have not adequately addressed the trend toward exploiting Web services (which, especially for enterprises offers several advantages in terms of code reusing, productivity and leveraging integration processes). A SOA is essentially a collection of services. These services communicate with each other. The communication can involve either simple data passing or it could involve two or more services coordinating some activity. Some authors (Vermeulen et al., 2007) propose to extend service descriptions with user interface information. For this purpose, the Web Services Description Language (WSDL) description is converted to Web Ontology Language for Services (OWL-S) format, which is combined with a hierarchical task model and a layout model. Model-driven design and deployment of service-enabled Web applications using Web Modeling Language (WebML) have been proposed as well (see, for example, (Manolescu et al., 2005). Usable security needs a different focus since it requires an environment based on HCI models for generating usable and secure service front ends which can be implemented in a variety of implementation environments and not only for the Web.

Since concrete logical descriptions are often specified through eXtensible Markup Language (XML) user interface languages (Luyten et al., 2004), there have been proposals to use XML tree algebra-based techniques (Lepreux et al., 2003), in order to compose entire or partial concrete presentations. The UI composition techniques are relevant in order to explore the possibility of composing existing UIs to access Web services. For example, the introduction of operations for combining interactors, such as fusion (composition with repetition of the intersection) and union (composition without repetition of the intersection) has been proposed (Lepreux et al., 2003). In that proposal, the general XML tree algebra (El Bekai & Rossiter, 2005) is applied to user interface composition and decomposition of graphical user interfaces specified in an XML-based language such as UsiXML (Limbourg et al., 2004). However, the method relies only on limited information and does not take into account the possible temporal constraints in the interactions.

Task models can provide useful support; in terms of granularity, tasks can be elementary tasks (namely, tasks that are considered as a logically atomic entity which cannot be further refined) or structured tasks (namely, tasks whose specification consists in decomposition and refinement of a number of logically-connected, smaller sub-tasks). It is also possible to identify task patterns: they are reusable structures in task models which can be used in various applications. Thus, whenever designers realize that the problem they are considering is similar to one that has already been found and solved, then they can reuse the solution previously developed. The hierarchical structure of this specification has two advantages: it provides a large range of granularity, allowing large and small task structures to be reused, and it enables reusable task structures to be defined at both a low and a high semantic level.

1.1.5.2 *User-Experience-Driven Design*

1.1.5.2.1 Service Engineering and Cloud Computing from a User Perspective

Miller (2008) lists the user benefits of cloud computing. With cloud computing, users do not have to take care of updating software to newer versions, but the newest version is always available. The user can have unlimited memory capacity on the cloud, and the data is automatically secured and backups are taken care of. With cloud solutions, users no longer face compatibility problems with different operating systems, different file formats or software versions. When a user's documents are stored in the cloud, the user has universal access to the data anywhere and always has the latest version at hand. The cloud service provider may also provide professional handling of security. User terminals can be more modest and cheaper than today as they need less local processing power and storage.

There are also threats related to cloud computing. Cloud computing relies on continuous, fast Internet connections. Those connections are not self-evident. In web-based applications, features may be limited compared to local applications (Miller, 2008). Security may become a threat if the user does not know where the data is stored, who has access to it, and how, where and for what purposes different data is combined. In principle, data in the cloud is safe and replicated across multiple machines. But if the data does go missing, the user will not have physical or local backup (Miller, 2008). Supporting trust toward the cloud services becomes a central user experience issue in the cloud. Also, data sanitation problems, profiling and identity thefts, backward compatibility as well as access right, privacy, liability, data ownership, and copyright issues can be seen as potential problems in cloud computing.

Today, many popular consumer services such as Facebook, Flickr, and email services are cloud-based. That is why consumers are used to this paradigm. However, users still commonly have local copies of their photos and other important documents, so the services are not yet totally cloud-based. As users have their local copies, they have not yet faced situations where, for example, photos stored in the cloud disappear because the company providing the service no longer exists. Problems related to data ownership have been foreseen but people have not yet commonly faced these problems in practice. Internet service providers are monitoring user behavior and collecting lots of user data without the users being aware of what is actually monitored. The cloud facilitates the manipulation and collection of users' data, and that can reveal very sensitive and personal issues of an individual user.

User experience (UX) is currently the most often used concept when studying new products and services from a user's point of view. UX emphasizes that products need to support users' hedonic needs such as stimulation and self-expression, in addition to the pragmatic ones (Hassenzahl and Tractinsky, 2006) in using the product or service. Designing for UX aims at broader views of users' emotional, contextual and dynamically evolving needs, and the impact of users' previous experiences to the new experiences. Furthermore, positive user experience means that the users' interactions with every contact point in the lifecycle of the system usage are satisfying, including taking it into use, active usage of the system, using the supporting services including maintenance and upgrading the system (Väänänen et al., 2008).

Väänänen et al. (2008) present a lifecycle framework for service user experience (SUX) that defines factors that need to be addressed to provide attractive and acceptable Web 2.0 services during the different lifecycle phases. Recently, ISO has been renewing the human-centered design standard so that it will include user experience in addition to usability. According to the renewed ISO 9241–210 Standards for Usability, Usability Reports and Usability Measures, user experience includes all the users' emotions, beliefs, preferences, perceptions, physical and psychological responses, behaviors, and accomplishments. User experience is a consequence of the presentation, functionality, system performance, interactive behavior, and assistive capabilities of the interactive system. It is also a consequence of the user's prior experiences, attitudes, skills, and personality. Rather than just a design target, user experience can be seen as a framework to study customers' attitudes toward a product and related services throughout the product lifecycle.

Both security and user experience are thought to be extremely important factors for the success of cloud computing (Chang, 2008; Cavoukian, 2008; Hayes, 2008; Rad et al., 2004). Security and user experience of cloud computing are strongly connected: most of the cloud security issues affect the user experience, but not all cloud user experience issues are security related. Trust is a concept that combines both these viewpoints, and building trust is one of the main user experience targets for any cloud service.

1.1.5.2.2 User Perception of Security

VTT Technical Research Centre of Finland Ltd. carried out a large online survey in June 2010 in Finland, USA, and Japan with over 3000 respondents. The survey studied cloud characteristics; it examined both the positive and the negative effects that the cloud brings both in the security point of view and user experience point of view. A large computer memory and processing resources, as well as pricing per use, were seen as important factors by most of the respondents. Also, the majority of the respondents stated the importance of the certainty that their documents are not copied or used elsewhere without their consent. The possibility for data sanitation, the reliable deletion of an uploaded document, was also agreed to be important by most of the respondents.

The possibility of someone stealing a user's identity on the Internet caused significant concern to the respondents. Privacy issues were also raised heavily; most of the people stated that they do not usually provide their contact details on the Internet. They also wanted to make sure that their personal information is not freely available on the Web.

Contrary to popular belief, information security was not considered boring or unnecessary; most of the respondents heavily stated that security is even more important to them than ease of use or good functioning of the service. It was also found that the novice users feel significantly more insecure toward the services in the cloud than the users who evaluate themselves to be on the intermediate or advanced skill level. Even though security is valued more important than ease of use or functionality, the security tools were still considered to annoyingly slow down the computer by relatively many users and answering security-related questions raised moderate frustration with some users. Hence, there is still a lot of improvement needed in the current security solutions.

From the user point of view, security should be transparent, automated, reliable, and always updated and available. Users are different; some want to have options to choose from, such as different levels of security and privacy settings; some neither understand nor want to understand security issues at all. Both kinds of users should have a safe and secure feeling about the cloud. Secured connection and good security level of a service or a website should be presented clearly. As mentioned, internet service providers are monitoring user behavior and collecting lots of user data without the users being aware of what is actually monitored (i.e., telemetrics). The cloud facilitates the manipulation and collection of users' data, and that can reveal very sensitive and personal issues of an individual user. Personal data scattered around the world in different servers creates a new privacy threat.

While security research has thus far focused a lot on access rights to services and data, in the future the need for tools where users can protect and safely and easily share their own content will increase.

1.1.5.3 Metrics-Based Usability and Security Evaluation

There are many models to evaluate the usability and security of the software, although there is no uniformity and consistency between these models. Moreover, there is no relationship between security and usability, and much less between its sub-characteristics.

With respect to usability, we can mention ISO/IEC 9126, which is one of the most widely used standards for evaluating the quality of a software product (ISO/IEC, 2001), and which defines a usability model of sub-characteristics as a component of software quality. The standard ISO 9241-11 (ISO/IEC, 1998) defines another quality model which identifies three attributes of usability (effectiveness, efficiency, and satisfaction). On the other hand, Systems and software Quality Requirements and Evaluation (SQuaRE) is a set of standards that interpret the quality of a software system as the degree to which the system satisfies the implicit and explicit needs of its different users (ISO/IEC, 25000). SQuaRE includes a significant evolution of the model proposed in ISO/IEC 9126. In the quality characteristics identified in the latest draft of the ISO/IEC 25010 standard, in accordance with (Bevan, 2010), usability is defined with some changes in the set of sub-characteristics of usability.

With respect to security models, ISO/IEC 9126, also makes a brief mention of the security as quality characteristic, but it defines it as a sub-characteristic of functionality (ISO/IEC, 2001). However, SQuaRE (ISO/IEC, 25000), defines security as a characteristic of quality, and it defines a set of sub-characteristics for security (confidentiality, integrity, non-repudiation, responsibility, authenticity, and conformity). It also defines the concept of availability and safety but detached from the concept of security. There are many other models that define security features, and, just as with the usability, in this case, also there are similarities and differences. Some of these models are proposed by Firesmith (2004) or may be derived from the framework for the assessment of product security known as Common Criteria (ISO/IEC, 2009), or can be extracted from MAGERIT (MAP, 2005), a methodology of analysis and risk management developed by the Higher Council for Electronic Administration of the Public Administrations of Spain, which is a reference document to both national and international level methodology officially recognized by NATO (CCN-CERT, 2008) and the OECD (OECD, 2005) for risk management, and conforms to various international security management, such as 13.335 (ISO/IEC, 2004). The family of ISO/IEC 27000 is also a reference in terms of security, and defines a model of security features, like COBIT (Control Objectives for Information and Related Technology), which represents a set of best practices for management of information security created by the Association for the Audit and Control Information Systems (ISACA).

Building a quality model based on security and usability not only improves the ability to evaluate in a systematic and unified manner these two properties but also enables reasoning about solutions that best integrate both quality properties.

1.1.5.3.1 Usability and Security Metrics

As mentioned previously, there is no uniformity in the definition of usability and security features of the software, which highlights the fact that there is neither no uniformity in the definition of metrics to evaluate these characteristics in the software development cycle. There are many proposals for the definition of security and usability metrics, but most of them are focused on the final product (in code), with very few metrics focused on the artifacts developed on the stage of analysis and design (Juristo, Moreno et al., 2007).

In particular, if we focus on usability, we can find some proposals that seek to identify metrics for analysis and design software. For example, in Folmer and Bosch (2003), the authors discuss the relationship between usability and software architecture. In Uldall-Espersen (2008), a set of informal questions on aspects of usability in software design are formulated. In Jokela, Koivumaa et al. (2006), techniques are proposed to assess whether the design of user interfaces meet the design requirements previously set. Finally, Seffah, Donyaee et al. (2006) propose a model to measure the usability called QUIM, which breaks down the usability into a set of factors, these factors in criteria and these criteria in metrics. Instead, there are much more usability metrics proposals related to the code or the developed software product. For example, proposals such as Bevan (1995); Hornbaek (2006); Seffah, Donyaee et al., (2006); Zou et al., (2007), and Zeman et al., (2009) measure the degree to which the software can be understood, learned, and operated attractively and in compliance with usability standards and guidelines.

As for security metrics, there are also proposals in the field of artifacts developed into the software analysis and design stages, but they have mostly focused on the final developed software product (i.e., code). Some of them are, for example, Alshammari, Fidge et al., (2009) who focus on the security of the design of object-oriented applications; Chandra, Khan et al. (2009) who provide a framework for the estimation of security in the design stage, or the Common Criteria or ISO/IEC 15408 (ISO/IEC, 2005), established a set of requirements to define security features of products and systems, information technology and the criteria to evaluate its security. On the other hand, the ISO/IEC 27004 (ISO/IEC, 2009) is a metric for security management. Other proposals, such as Common Weakness Enumeration (CWE) (Martin, 2010), provide a unified set of software vulnerabilities that make measurable and effective discussion, description, selection, and use of security tools and services software, which is related to Common Vulnerability Scoring System (CVSS) (Mell and Scarfone, 2007), which provides recommendations for scoring vulnerabilities in the design, and Common Misuse Scoring System (CMSS) (Ruitenbeek and Scarfone, 2009), a scoring scheme to measure the severity of the vulnerabilities of misuse of software features.

Other proposals, such as NIST 800-55 (Swanson, Bartol et al., 2003) and Wang & Wang (2009), provide guidelines for designing security metrics. The number of identified security metrics which are applicable to the code is much higher. These proposals are defined not only in the scientific community but also for professionals. Some websites can be mentioned, such as www.securitymetrics.org or https://www.metricscenter.net, which provide a huge set of security metrics over the code. Other proposals, such as Chowdhury and Zulkernine (2010), present a framework to automatically predict vulnerabilities, and others, like McGraw (2008), summarize the code review tools for security. In addition, some authors propose frameworks to drive security improvement through metrics, as in Nichols and Peterson (2007). In Pfleeger (2009), the authors identify the characteristics they consider key to distinguishing between different security entities (systems, organizations, companies, business or industry). Other authors also discuss issues like complexity, version and developer activity as an indicator of software vulnerabilities (Shin, Meneely et al., 2010), and there are further proposals which define security metrics for specific software domains, as is the case of Walden & Doyle (2009), which is specific for web application security. In conclusion, we consider that there is now a multitude of proposals related to source code analysis and detection of vulnerabilities. Also, so far there is great instability in the definition of security metrics over the code, which highlights that security measurement is sometimes ambiguous and imprecise, so the measurements can vary depending on the interpretation of the metrics.

1.1.5.4 Usable Secure Design Patterns and Pattern-Oriented Design

In his two books, *A Pattern Language* (Alexander et al., 1977) and *A Timeless Way of Building* (Alexander, 1979), the architect Christopher Alexander ("the father of patterns") introduced the concept of design patterns as a "three-part rule that expresses a relation between a certain context, a problem, and a solution" (Alexander, 1979). Following the object-oriented software design community (Gamma et al., 1995), software practitioners investigated design patterns as one possible solution to reuse design knowledge via class libraries. However, design patterns can or should think and be applied at higher levels than mere code. Patterns can also be seen as a vehicle to reuse successful practices, to reason about what is done and why in terms of design solutions, and to disseminate the solutions. They are a useful tool for dealing with multidisciplinary problems such as the security and usability engineering ones.

During the last 15 years, the HCI community has also adopted patterns as a user interface design tool. Tidwell (1997) describes a user interface pattern as possible good solutions to a common design problem within a certain context, by describing the invariant qualities of all those solutions. Using patterns should be like asking your experienced colleague in the next room for advice (Granlund and Lafrenière, 1999). For the problem of the conflict between usability and security, a secure usable design pattern can be seen as a proven solution to a problem that occurs in several contexts.

The origin dilemma of the usability/security problem can be summarized as follows. The most common developers are not trained in security or usability nor are they experts in both. In industry, very few teams have a security specialist, a usability specialist or both, which would be the ideal solution. Given this evidence, the number of teams that have high expertise in both security and usability is likely to be small. Our goal is to investigate patterns as a solution to this situation via the promotion of well-known security usability problems related to Web and cloud services.

A number of pattern languages have been suggested in HCI. For example, (Duyne et al., 2003) "The Design of Sites", (Welie, 2008) Interaction Design Patterns, and (Tidwell, 1997) UI Patterns and Techniques play an important role. In addition, specific languages such as user interface design patterns (Laakso, 2003) and the UPADE Language (Engelberg and Seffah, 2002) have been proposed as well. Different pattern collections have been published, including patterns for Web page layout design (Tidwell, 1997) and (Coram and Lee, 1998) for navigation in large information architectures, as well as for visualizing and presenting information. These patterns do not address security even when it is listed as a sub-factor that contributes to the overall usability.

Various security design patterns are available online (http://www.securitypatterns.org/patterns.html, http://www.scrypt.net/~celer/securitypatterns). Recently, many practitioners suggest also security patterns for Web services (http://coresecuritypatterns.com/patterns.htm). Again, this catalogue and other similar ones are pattern-related to only the security concerns of technology. They are also more related to the code and programming language rather high-level conceptual solutions.

By identifying and disseminating a new generation of design patterns that bridge the security and usability of existing design patterns languages, it is reasonable to expect a kind of harmony between usability and security, as well to deal with the new challenges of tomorrow's secure usable services. Beside identifying and validating a catalogue of patterns, a tool to support the design and testing of secure usable service with patterns needs to be designed. Not only does usability and security testing come late in the design process, but also testing is expensive and time-consuming, and results are rough and mainly qualitative, often not providing sufficient insights about the cause or solution of a problem. Due to the expense and difficulty of testing, often it is not possible to optimize a design means of enhancing its usability. In addition, this difficulty makes specialists reluctant to try innovative but risky designs. What is needed is a cost-effective vehicle for capturing and disseminating the best-design practices. It takes also an integrative model that can predict how users would perform in real life, thereby giving designers immediate feedback on their design in progress and the ability to experiment easily with different design alternatives.

Patterns can be as a vehicle for integrating the best secure usable design solutions into our human-centric framework and into the service engineering lifecycle. Most design pattern languages and pattern-oriented design frameworks include little information about end-users. When they do, they are hidden among paragraphs of textual information contained within pattern descriptions. For example, the "redundant encoding" pattern indicates among its context that users may have visual deficiencies such as color blindness (see www.welie.com). Another example is the "Wizard" pattern that gives guidance to users. It is important for the Web services designers to know what kind of added value patterns bring, especially to their users and in terms of usability and security quality attributes and measures. This kind of information should not simply be hidden in pattern descriptions but should be part of the elements that make up a pattern. Therefore, pattern-based designs should be considered as design blocks that apply in particular situations, derived from an understanding of user experiences with security systems.

The following is an example of a basic usable secure design pattern:
Website Password Prompts (Source Ka-Ping Yee, 2002)

PROBLEM

Suppose that Alice and Bob both run websites requiring user authentication. Both use the same free hosting service at example.org. If we open two browser windows, one at each site, and attempt to enter the protected areas, two password prompts will appear (see the following figure).

SOLUTION

A possible design that would solve both of these problems consists of first introducing the rule that applications are only allowed to draw into rectangular frame, which the system copies onto the screen (Figure 1.2). The system manages the window borders and the desktop area. Then, we change the Web browser so it asks the operating system to request user authentication on its behalf. The system-generated password prompt is drawn with a red-striped border that no application could imitate, eliminating the possibility of spoofing. The prompt could even be animated or faded in to demonstrate the system's exclusive ability to draw anywhere. Lines join the prompt window to the window of the requesting application, establishing an unmistakable association.

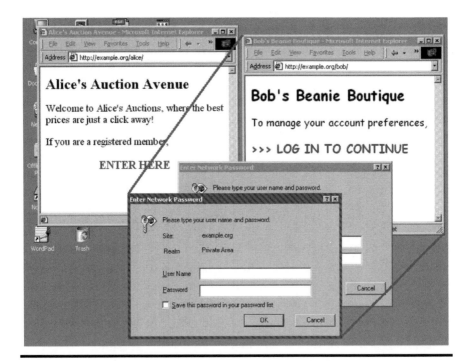

Figure 1.2   A possible usability solution to securely manage an identity problem.

Chapter 2

Panoramic Overview of User Authentication Techniques

Security systems are conceived to allow authorized users in and to keep unauthorized users out of an organization's network resources. In addition, the security system needs to make sure that users only perform actions they are authorized to perform. To this end, user authentication is the entry point to different computing networks or facilities in which a set of services are rendered to users or a set of tasks can be performed. To be authenticated, users need to *log in* (also, to log on, sign in, or sign on) to identify themselves to the system in order to obtain access. On the other hand, users also need to *log out* (also, to log off, sign out, or sign off) to close off their access to a computer system after previously logging in. For example, once successfully authenticated, the user can gain access to a company's Intranet, databases, applications, facilities, etc.

Why focus on authentication? Authentication is more important than ever due to the collapse of network security perimeters, the expansion in the number of devices wanting to access company networks, and the rising number of remote users and wireless devices (including laptops) since users want to access increasingly diverse applications. The information that users (i.e. average corporate/consumer computer users) need to access has broadened to comprise all aspects of both personal and business purposes, including email, a greater range of applications, and various types of data. In particular, there has been an impressive increase in corporate users' need to access their organizations' network resources, characterized by a growing number and variety of users (e.g. local and mobile users, telecommuters, etc.), applications, access methods, and extensions of enterprise networks to include third parties (i.e. customers, suppliers, partners, employees, consumers, etc.), hence exposing organizations to significant risks unless they take protective measures.

Gartner Research (Ng et al., 2018) predicts spending in public cloud services will increase by 22.9% to $186.9 billion in 2018. Also, cloud-delivered authentication services continue to grow faster than the overall growth for the user authentication market, and Gartner expects that this will extend, as "multitenanted" services mature and as cloud becomes increasingly adopted as a more effective way of delivering any application or service. This is also resonated in the increasing usage of cloud-delivered authentication methods tools and the increasing use of those tools to meet user authentication needs in cloud-forward enterprises (Market Guide for User Authentication, published 16 November 2017).

Although (strong) multi-factor authentication (MFA) surely enhances security, there is no guarantee that end users will adopt it as a convenient and usable authentication factor. "Reflecting the real-world difficulties security managers have had in keeping users happy with the choice of strong authentication, the vendor landscape is complex and fragmented. Vendors will continue to expand their product lines until enterprise adoption of identity management, including strong authentication, becomes more widespread in the coming years" as shown in Forrester's Enterprise and SMB Security Survey, North America And Europe, Q3 2007. Additionally, an organization's authentication service should be suitable to the risks, and it should consider the impact on users, as well as the cost of integration with its existing technology architecture, and total cost of ownership.

2.1 The Context of Authentication in Computer Security

Identification, authentication, and authorization are distinct and necessary components that allow users to securely access a computer system. Authentication is the process of establishing whether someone is who s/he declares her/himself to be. In private and public computer networks (encompassing the Internet), authentication is popularly done through the use of logon passwords. The logon is the process used to gain access to an operating system (OS) or application, generally from a remote computer. Usually, a logon requires that the user have a user ID (username) and a password.

Authentication is one of the critical elements of a set of services that constitute a security sub-system in a communications infrastructure and encompasses the following security services:

- *Authentication*: The verification of a claimed identity.
- *Confidentiality*: The property that information is not made available or disclosed to unauthorized individuals, entities, or processes.

- *Integrity*: The property that data has not been modified or destroyed in an unauthorized manner.
- *Non-repudiation*: The process of ensuring that the author of a document cannot later claim not to be the author.
- *Access Control*: Encompasses any mechanism granting access to data or performing an action. An authentication method is used to check a user login; then the access control mechanism grants and revokes privileges based on predefined rules.
- *Availability*: Demands that a computer system's assets be available to authorized parties when needed.

The three essential security properties of *confidentiality, integrity,* and *availability* rely on the differentiation between authorized and unauthorized users. In order to differentiate between them, authentication must be present. The authentication process is based on a risk criterion. High-level risk systems, applications, and information necessitate distinct forms of authentication that more precisely affirm the user's identity than would a low-level risk application, where the confirmation of the identity is not as significant from a risk standpoint (e.g. anonymous authentication in a library). This is typically referred to as "stronger authentication".

The authentication services are located beneath the security operations ring in an ideal organization's computer security framework (Figure 2.1). The trusted security framework is no longer a limited technological matter: It supports strategic initiatives and provides a platform for taking a business to new levels of competitiveness. The enterprise security framework must provide confidentiality, integrity, and availability throughout the enterprise, bringing it into conformity with corporate objectives.

The framework is centered around the business assets to be protected, which are identified and prioritized through the security strategy and management. The security management is in charge of coordinating and supervising the different security services (security operations, security compliance, security policy and standards, and, finally, security awareness). Security services are the services supplied by a system for implementing the security policy of an organization. A standard set of such services includes identification and authentication, access control and authorization, accountability and auditing, data confidentiality, data integrity and recovery, data exchange, object reuse, non-repudiation, and, finally, reliability.

2.2 User Authentication Market

According to the latest Gartner's Magic Quadrant for User Authentication (Allan et al., 2014), (Figure 2.2) a vendor in the user authentication market delivers on-premises software/hardware or a cloud-based service that makes real-time authentication

Figure 2.1 User authentication in an ideal organization's security framework (Accenture, 2004).

decisions for users who are using an arbitrary endpoint device (i.e., not just Windows PCs) to access one or more applications, systems, or services in a variety of use cases. The market is dominated by 10% of the authentication vendors globally. While mobile and cloud remain disruptive, buyers continue to give weight to user experience. In Best Practices for Selecting New User Authentication Methods (Gartner Research, 2 February 2018), user wants and needs dictate the overall level of UX, and indicate which specific aspects of UX are more or less important: i) Poor UX can drive compensating user behaviors that erode trust and accountability, so good UX is a way of ensuring a desired level of risk mitigation; ii) Poor UX can drive up training and support costs, increasing TCO. Investment in contextual, adaptive techniques increases as well as in biometric methods. Smart things will become authenticators.

Where appropriate to the authentication methods supported, a vendor in this market also delivers client-side software or hardware that end users utilize to make those real-time authentication decisions.

The market is mature, with several vendors offering products that have been continuously offered during the past three decades (although ownership has changed over that time). However, new methods and vendors continue to emerge, with the most rapid growth occurring within the past decade in response to the changing market needs for different trade-offs among trust, user experience (UX), and total cost of ownership (TCO). The greater adoption of user authentication over a wider

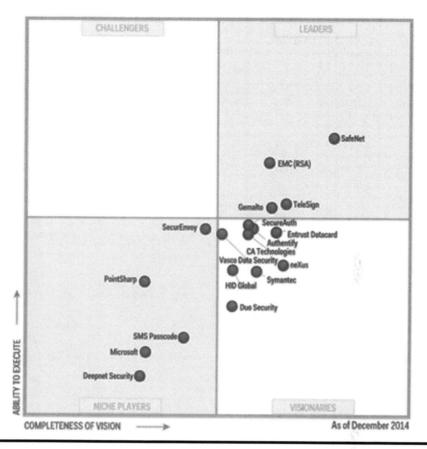

Figure 2.2   Magic quadrant for user authentication (Allan et al., 2014).

variety of use cases, the impact of mobile, cloud and big data analytics, and the emergence of innovative methods continue to be disruptive. Gartner is aware of more than 250 vendors offering some kind of stand-alone user authentication product or service, although only approximately 100 of these might be commercially viable, and perhaps fewer than 50 vendors have offerings that we would consider to be credible choices for Gartner clients.

This Magic Quadrant research covers the 18 vendors with the most significant market presence by number of customers or number of end users served (see the Inclusion and Exclusion Criteria section), although these numbers vary by orders of magnitude. The largest vendors in this Magic Quadrant account for the majority of the market by customer and end user numbers. Gartner defines "user authentication" as the real-time corroboration of a claimed identity with an implied or notional level of trust. This is a foundational Identity and Access Management (IAM) function, because without sufficient confidence in users' identities, the value of other IAM functions – such as authorization (especially segregation of duties), audit, and analytics – is eroded.

2.3 User Authentication Use Cases

Vendors included in the Magic Quadrant (Allan, et al., 2014) typically can support multiple use cases. However, not all vendors have equal experience in all use cases; some may have a stronger track record in use cases such as workforce remote access while others may focus on access to retail customer applications, especially in financial services, either of which might limit their vertical position within the Magic Quadrant. Gartner analysis of the vendors' market responsiveness and track record considered (among other things) each vendor's demonstrated ability to support organizations' needs across the variety of use cases enumerated in the following.

2.3.1 Endpoint Access

- PC pre-boot authentication: Pre-boot access to a stand-alone or networked PC by any user.
- PC login: Access to a stand-alone PC by any user.
- Mobile device login: Access to a mobile device by any user.

2.3.2 Workforce Local Access

- Windows local area network (LAN): Access to the Windows network by any workforce user.
- Business application: Access to any individual business applications (Web or legacy) by any workforce user.
- Cloud applications: Access to cloud applications, such as salesforce.com and Google Apps, by any remote or mobile workforce user.
- Server (system administrator): Access to a server (or similar) by a system administrator (or similar).
- Network infrastructure (network administrator): Access to firewalls, routers, switches, and so on by a network administrator (or similar) on the corporate network.

2.3.3 Workforce Remote Access

- Virtual private network (VPN): Access to back end applications via Internet Protocol Security (IPsec) or Secure Socket Layer (SSL) by any business partner, supply chain partner, or other external user.
- Hosted virtual desktops (HVD): Access to the corporate network via a Web-based thin client (e.g. Citrix XenDesktop or VMware Horizon View) or zero client (e.g. Teradici) by any business partner, supply chain partner, or other external user.
- Business Web applications: Access to Web applications by any business partner, supply chain partner, or other external user.

- Portals: Access to portal applications, such as Outlook Web App and self-service HR portals, by any remote or mobile workforce user.
- Cloud applications: Access to cloud applications, such as salesforce.com and Google Apps, by any remote or mobile workforce user.

2.3.4 External Users' Remote Access

- Virtual private network (VPN): Access to back end applications via IPsec or SSL-VPN by any business partner, supply chain partner, or other external user.
- Hosted virtual desktops (HVD): Access to the corporate network via a Web-based thin client (e.g. Citrix XenDesktop or VMware Horizon View) or zero client (e.g. Teradici) by any business partner, supply chain partner, or other external user.
- Business Web applications: Access to Web applications by any business partner, supply chain partner, or other external user (except retail customers).
- Retail customer applications: Access to customer-facing Web applications.

2.4 Elements of User Authentication

In an authentication process there are some elements that are present as shown in Figure 2.3:

- A user (or principal) to be authenticated. The principal is the entity requesting authorization. It is generally some combination of user, device, and/or service (e.g. log in to the computer system).
- A credential, which is possessed by the user who submits it as proof of identity. The main types of credentials are shared key (password), one-time password (OTP), digital certificate, and biometric credential.

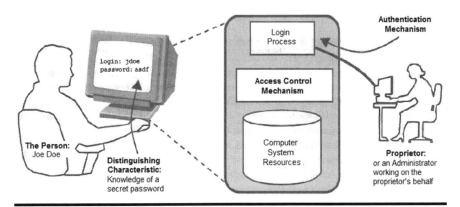

Figure 2.3   The elements of the user authentication process (Smith, 2002).

- A distinguishing characteristic that sets apart that particular user.
- A proprietor who is responsible for the system in use.
- An authentication mechanism to verify the existence of the distinguishing characteristic.
- A server, which is the authentication key storage.
- A privilege when the authentication is successful by employing an access control mechanism, which rejects the privilege if authentication is unsuccessful.
- A contextual information of the authentication request that encompasses the network and physical location of the request (e.g. geolocation, IP address, workstation), the kind of access provided (e.g. check balance), the time of day, and other elements such as network load, security threat level, and so on.

2.5 Architectural Design Patterns in Authentication

There are some architectural design patterns which have frequently been encountered in the deployment of authentication systems (Smith, 2002), such as local authentication, direct authentication, broker authentication, and offline authentication. These patterns analyze problems in terms of space and people, not data and processing. The following is a description of the architectural design patterns employed in authentication:

- Local authentication: This is related to single desktop systems and laptops. The whole system (comprising its authentication and access control mechanism) lives within a single physical security perimeter.
- Direct authentication: Figure 2.4 shows the direct authentication when a client and service share a trust relationship. Direct authentication requires the presentation of credentials, which are usually a username and password. The service employs these credentials to authenticate the request.

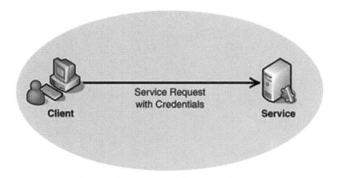

Figure 2.4 Direct authentication when a client and service share a trust relationship (Microsoft, 2005).

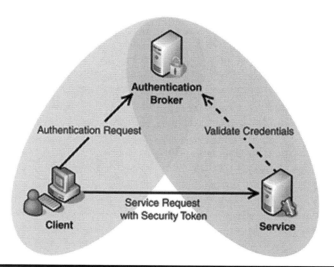

Figure 2.5   Broker authentication (Microsoft, 2005).

■ Broker authentication: Using a broker to carry out authentication when client and service do not share a trust relationship, as shown in Figure 2.5. The broker authenticates the client and then issues a security token that the service can employ to authenticate the client. The security token is constantly verified, but usually the service does not need to interact with the broker to carry out the verification. The reason for this is that the token itself can include proof of a relationship with the broker, which can be employed by the service to verify the token.

■ Offline authentication: Public key certification software follows an offline authentication pattern which recognizes authorized users, is stored in multiple locations throughout the system, and is accessible offline. For example, if a legitimate user wants to authenticate to her bank's server, her workstation first acquires the bank's public key certificate and authenticates it with a pre-established public key. Second, it uses the public key within the bank's certificate as part of some other protocol, such as SSL, to authenticate the bank as the real owner of the private key, which is mathematically related to the certificate's public key. (Public key authentication is described later in this chapter.)

2.6 Authentication Factors

In some applications, there is no need for users to be authenticated by the system, as in the case of browsing the Internet in a public library. So, if the data is public, there is no need to limit access to users or, more explicitly, no need for access control. If no authentication is taking place, the user is said to be "anonymous". This is the initial status after a connection has been opened to the server. However, when the resources are protected, for example in an online banking website, the user must be authenticated in order to have access to the services. The fundamental purpose of security is to control who has access to valuable property, whether physical or logical. An authentication factor is a piece of information used to authenticate or verify a person's identity.

There are four factors of user authentication that might be employed in combination to increase the level of security in the claimed identity of a user according to Table 2.1:

Table 2.1   Guide to Understanding Identification and Authentication in Trusted Systems (NCSC, 1983)

Classification	Factor	Examples
Type 1: Authentication by Knowledge	Something you *Know*	• A password or passphrase[a] • A Personal Identification Number (PIN). • Information about the user or family members
Type 2: Authentication by Ownership	Something you *Have*	• A physical key. • A magnetic stripe card. • A token that generates a one-time password (OTP).
Type 3: Authentication by Characteristic	Something you *Are* (or a physical attribute)	A Biometric trait: • Fingerprint. • Iris pattern. • Hand geometry. • Voice.
Type 4[b]: Authentication by Emanation (Braz and Aïmeur, 2003)	Something you *Convey*	A microprocessor-chip computer (ChipTag) implanted under human skin. This ChipTag is able to authenticate users' access to systems and connect them wirelessly through radio frequency identification (RFID) (Braz and Aïmeur, 2003)
Or a combination of the aforementioned		

[a] Passphrase is generally longer than a password and includes letters, numbers, words, and random characters. In encrypted communications, one should always use a passphrase rather than a password. For example: *I must g!o down to the sea again, to t7he lonely s8a and the s1y.*

[b] This is a novel (fourth) authentication factor that has been proposed by Braz and Aïmeur, (2003).

Figure 2.6   RSA SecurID® 700 hardware authenticator with One-Time Password (OTP).

- Something you have (e.g. smart card).
- Something you know (e.g. password or PIN).
- Something you are (e.g. iris recognition).
- Something you convey (e.g. mobile radio frequency identification [RFID] implantable chip: a ChipTag embedded with a unique ID that is implanted beneath the skin of a user's upper-arm. This user ID is conveyed through radio frequency [RF] signal to the mobile device antenna to authenticate the user to the system. This novel authentication method also introduces a novel [fourth] authentication factor that has been proposed by Braz and Aïmeur [2003]).

Using any authentication factor alone provides single-factor authentication. Any two authentication factors may be combined to provide two-factor authentication; NCSC-TG-017 calls these combinations Type 12 ("one-two"), Type 13 ("one-three"), and Type 23 ("two-three") as shown in Table 2.1. Associating all three factors presents three-factor authentication, Type 123 ("one-two-three"). So, associating two or more factors introduces greater security. The most well-known example is the magnetic stripe card (smart card) and PIN (Two-Factor Authentication) used with an Automated Teller Machine (ATM). To access the network, a legitimate user must have both "factors", just as he must have an ATM card and a PIN to withdraw money from a bank account. Another example is the RSA SecurID® 700* hardware authenticator (Figure 2.6), which enforces strong "two-factor" authentication by requiring the user to present two forms of credentials to prove his identity: something you know (a password or PIN), and something you have (authenticator). An example of a multi-factor authentication method is the combination of password, smart card, and iris recognition, resulting in far greater security (e.g. it can be employed for higher risk transactions).

As mentioned, when the resources are protected, the user must be authenticated in order to have access to the services (i.e. access control). The purpose of access control is to limit the actions or operations that a legitimate user of a computer system can perform. Access control constrains what a user can do directly,

---

* RSA SecurID® Hardware Tokens form RSA Security. Source: https://www.rsa.com/content/dam/en/data-sheet/rsa-securid-hardware-tokens.pdf

as well as what programs executing on behalf of the users are allowed to do. In this way, access control seeks to prevent activity that could lead to a breach of security. It is important to make a clear distinction between authentication and access control. Correctly establishing the identity of the user is the responsibility of the authentication service. Access control assumes that the authentication of the user has been successfully verified prior to enforcement of access control via a reference monitor.

Consider the following contextual authentication scenario: Before a system grants a legitimate user access to a company's protected network resources, the system must determine who she is, if she belongs to the system, if she has the right to access the system, and if she is the person she says she is. Actually, the system has required three distinct elements – *identification, authentication, and authorization* – that together comprise the so-called *access control*. However, how does the system confirm that she is who she says she is? For example, entering her password does not prove it is her. Hence, the system needs the identification and authentication to authorize access for her. The authentication information may be gathered from one of the authentication factors shown in Table 2.1.

As already mentioned, authenticated identities are the foundation for several other information security services. Typically, an organization needs to do the following:

- ■ Control individual users' access to its information systems (Authentication);
- ■ Control individual users access to the resources and services supplied by those systems (Authorization);
- ■ Generate an audit trail of users' access or attempted access to those systems, resources, and services.

2.7 User Authentication Methods

This section describes the most representative user authentication methods currently available, such as passwords and PINs, authentication tokens (also known as software or hardware authenticators), mobile phone as a token authentication method, Kerberos, biometrics, and single sign-on (SSO). It also introduces an advanced user authentication method called under-skin RFID chip.

2.7.1 Passwords and Personal Identification Numbers (PINs)

To avoid other users from using your account through your username, you are required to have a password. A username is a unique identifier, which can be system-defined or user-defined. There are several types of passwords such as weak, strong, system-defined, or user-defined passwords, passphrases, and PINs. Passwords, passphrases, and PINs are considered knowledge-based authentication (KBA) methods.

KBA is an authentication scheme in which the user is asked to answer at least one "secret" question. KBA is often used as a component in multi-factor authentication (MFA) and for self-service password retrieval.

Usernames are usually built of text so that people can easily remember and type in their usernames during the logon process. A password is a secret word or string of characters that is employed for authentication to prove identity or gain access to a resource. A password allows you (and only you) to access protected network resources in a computer system. A simple password such as hello, which is, in fact, easy to remember, has weak password strength, while a complex password such as hCytW7m9!, which is hard to remember, has strong password strength. *Password strength* is a measure of the effectiveness of a password in resisting guessing and brute-force attacks. The strength of a password is a function of length, complexity, and unpredictability. A password essentially proves to the computer system that you are who you say you are.

Another method for solving the problems intrinsic to picking passwords that are both easy to remember and hard to guess is cognitive passwords. This technique asks the user a series of questions that are easier for them to answer than for others. However, there are security issues related to this method if people close to the user – especially a spouse – could successfully answer many of the questions. Thus, this technique doesn't appear viable.

A long password, especially one with inserted spaces, is called a *passphrase*. A passphrase is a special type of token-based password where the tokens are words instead of symbols from a character set (e.g. "Where is my checked shirt?"). In principle, the longer the password the stronger it is. (In fact, it is less guessed and less exposed to certain kinds of attacks). The use of the passwords is by far the most common KBA (Type 1) user authentication method (Figure 2.7).

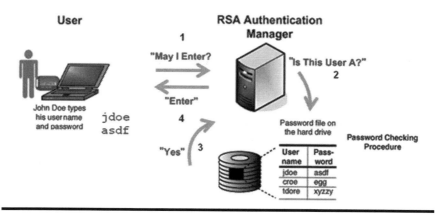

Figure 2.7 **The password authentication process (adapted from Smith (2002) and RSA Authentication Manager 8.3, https://community.rsa.com/community/products/securid/authentication-manager-83.)**

The problem with passwords is that if they are not encrypted, or if the encryption is easy to break, passwords and passcodes (i.e. a PIN plus a token code) are vulnerable to eavesdropping and replay. And even if it is encrypted, there are other types of attacks that may be used, such as a "brute-force" (or dictionary) attack, which consists of an attack that just tries possibility after possibility until the right one is found. Utilities to help an attacker with this kind of attempt are easily found on the Internet. Short passwords, made of one simple word, are the easiest to figure out with this kind of attack. Therefore, many Super Administrators require passphrases, complex combinations of words. Controls will also often require the use of a password policy as mentioned above, which makes it more random in nature and harder to guess. In some environments, users must remember many complex passwords and passphrases and end up writing them down near the computer. This becomes the vulnerability (EMC, 2009).

Apart from passwords, PINs are another form of user authentication. A PIN is a unique personal character string used as a password, usually with a four-digit number, which must be entered by the user before a remote terminal (e.g. desktop, mobile device, ATM, etc.) or point-of-sale terminal (e.g. kiosk) and can be used to transfer information or complete a transaction. PINs are often employed with a magnetic stripe card or smart cards at an ATM to authenticate, for example, a bank customer.

A classic strategy to defend against PIN guessing attacks in authentication tokens is to lock a user's account usually upon three consecutive invalid PIN attempts. Since the PIN is locked, it can only be unlocked by the help desk administrator. Until then, the user will be not able to logon to the system. SafeNet tokens*, for instance, will only generate the correct password after the correct PIN has been entered, and it has attack lockouts if the wrong PIN is entered too many times. In the worst-case scenario, the token becomes totally blocked and service is not available.

2.7.2 Security Questions

Security questions are a KBA user method to provide users with a secondary method to identify themselves online or when, for example, contacting the customer services of an organization to update your personal information, etc. Security questions are designed to be memorable to users but hard for anyone else to guess. When used in conjunction with other identifying information, they help the system's owner to verify that a user is the person requesting access to his account.

A good security question has the following characteristics:

- Easy to remember, even five or ten years from now.
- As a minimum, thousands of possible answers.

---

* SafeNet OTP 110 from Gemalto. Source: https://safenet.gemalto.com/multi-factor-authentication/authenticators/safenet-otp-110/.

- Not a question you would answer, for example, on Facebook in a "Fun Questions to Ask" survey, or in an article or interview.
- Simple one- or two-word answer.
- Never changes.

Some items to avoid when creating security questions:

- Favorite food, colors, etc.: these change over time.
- Vehicle make and model: there are only so many types of cars, trucks, etc. Most people could rattle off the popular makes and models of an era rather easily.
- Birthdays: birthdays are poor because they are easy to find online, even siblings or parents, since most social networking sites will send out alerts to everyone when birthdays are approaching.
- Name or birthday of a family member: if they are family there is a good chance they are your friend on a social networking site, so this information would be easy to find.
- School name, location, etc.: it is often easy for someone to find out the area a person lives or grew up, and there is often only so many schools in an area.
- First job location, name, etc.: it is often easy to find out where someone grew up, and there is a limited number of popular first jobs.
- The color of a particular object or item: Probably even friends and family can know the color of something, and there might even be photos of your vehicle on your Instagram, Facebook, etc.

2.7.3 Authentication (Security) Tokens

An authentication token (*also called a hardware token, authentication token, USB token, cryptographic token, software token, virtual token, or key fob*) may be a physical device that an authorized user of computer services (e.g. security administrator) has given to specific users to prove their identities. To verify the identity of a token's owner, the host system performs its authentication protocol using data encoded on the token. Some authentication tokens contain advanced components like a microprocessor and semi-conductor memory*, and they support sophisticated authentication protocols, which provide a high level of security. Generally, authentication tokens allow the use of SSO systems that enable users to utilize an authentication token to sign in only once to a wide range of applications or websites for which they demand access.

---

* RAM and ROM are semi-conductor memories. One of the characteristics of the semi-conductor memory is the ability to write information at an extremely high speed and read it out.

Authentication tokens come in a variety of physical forms. The size, shape, and materials from which a token is manufactured are referred to conjointly as the token's *form factor*. A token's form factor involves trade-offs that must be evaluated for deployment in different applications so security professionals can choose the form factor that is best suited to a particular system or application. An authentication token can have different form factors depending on the authentication method and vendor. Some of current and popular authentication tokens are

- Hardware token
- Software token
- Universal Serial Bus (USB) token
- Cryptographic token
- Virtual token
- Smartphone device with touch keyboard or physical keyboard

There are two physical authentication token types: disconnected tokens and connected tokens.

2.7.3.1 Disconnected Tokens

Disconnected tokens have neither a physical nor a logical connection to the client computer; nor do they, like proximity cards, have physical contact with a token reader device (e.g. an employee brings the proximity card close to the card reader to gain physical access into the office). They typically do not require a special input device, and they use a built-in screen to display the generated authentication data, which the user enters manually themselves via a touchscreen, keyboard, or keypad, as shown in Figure 2.8. They are the most common type of security token used

Figure 2.8 Symantec Validation & ID Protection (VIP) Access for Mobile. (https://m.vip.symantec.com/learnmore.v.)

(usually in combination with a password) in two-factor authentication for online identification.

The following are disconnected tokens types, which will be described in the next sections: One-Time Passwords (OTP), Challenge–Response (C/R) Authentication, and Mobile-Phone-As-A-Token Authentication Method.

2.7.3.1.1 One-Time Passwords (OTP)

As a security system that requires a new password every time users authenticate to a system, OTP generators make it more difficult to gain unauthorized access to restricted resources, such as by a hacker replaying an intercepted password. An OTP system usually employs microprocessor-based authentication tokens, which do not demand a physical connection to host systems. These devices communicate directly with users through an embedded display and some form of keypad. Users transmit authentication data, such as passwords or encrypted challenges, between tokens and host systems manually. Examples of other OTP generators are the following: SafeNet OTP 110 RemoteAccess (Figure 2.9), RSA SecurID® Software Token 1.1 for iPhone Devices (Figure 2.10), and DigiPass from Vasco (Figure 2.11).

OTP generators produce a sequence of passwords that are synchronized with host systems. Each password is only valid for one authentication, and so it cannot be recorded and replayed to obtain access. Synchronization is frequently based on a secret initial source value, which is swapped at specific time intervals or each time an authentication event takes place. Then, without knowledge of the secret value and the number of times it has been swapped, an attacker may not foresee the next password in the sequence even if one or more previous passwords are known. When an OTP is combined with a PIN, two-factor authentication is achieved because the client needs to have something (the token) and know something (the PIN). But how exactly does authentication via OTP work? Figure 2.12 shows the OTP authentication process.

1. User opens and connects to the VPN application from his computer.
2. User enters his username (jdoe) and a 4-digit numeric PIN (7234) in the username and passcode fields on the login application screen.
3. User types and appends the six-digit numeric token number (currently displayed on her token) to the entered PIN on the login screen. This number (7234435961) becomes the *passcode,* which is, in fact, the combination of PIN and token number.
4. User clicks "OK".
5. If the authentication is successful, the authentication server (AS) validates the passcode and grants user access to the network's protected resources.

Figure 2.9 SafeNet OTP (One-Time Password) Authentication. (From Gemalto, https://safenet.gemalto.com/multi-factor-authentication/authenticators/safenet-otp-110/.)

Figure 2.10 RSA SecurID® for iPhones. (From RSA Security, https://itunes.apple.com/ca/app/rsa-securid-software-token/id318038618?mt=8.)

Figure 2.11 DigiPassGo3. (From Vasco, https://www.vasco.com/products/two-factor-authenticators/hardware/one-button/digipass-go-3.html.)

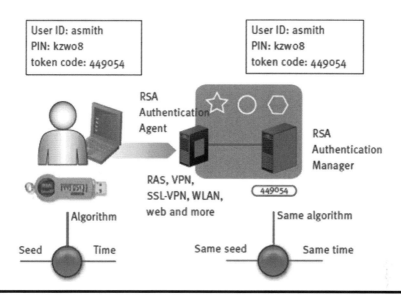

Figure 2.12 Authentication via OTP: Pre-condition: Authentication server randomly generates OTPs (six-digit numeric token numbers) on the user's token display (e.g. 435961) every 60 seconds. User has an assigned authentication token. (Enterprise-class security engine for RSA SecurID authentication, https://community.rsa.com/docs/DOC-62315.)

Limitations of OTP. OTP had been initially created to enhance the security of static passwords, and they were considered cutting-edge practically three decades ago. Nowadays, financial institutions continue to rely on this OTP-based authentication system for online transactions and authentication security, ignoring their extensively reported vulnerabilities.

Regardless of the OTP being used, all these methods share common flaws and vulnerabilities such as:

- Each of these methods is completely symmetric – the bank has access to the exact same information its customers do.
- All OTP systems continue to trust in browser-based communications back to the bank. For that reason, if fraudsters were to set up a phishing site to impersonate the bank's own online portal or the browser was someway compromised, user credentials and the OTP can be effortlessly captured by the attackers and instantly put to use to gain access to accounts and authenticate a fraudulent transaction.
- If using methods of sending OTPs through SMS channel, this is considered insecure for a large number of reasons:
  - The security of the SMS channel relies mainly on the security parameters of the cellular networks, and without access to Global System for Mobile

Communications (GSM) or 4G networks, the security of text messages cannot always be supposed.

– Today's mobile devices are extremely vulnerable to Trojan viruses, such as Zeus, Perkal, Citadel, and Zitmo, which capitalize on open access to the SMS channel on mobile devices to capture OTPs. Fraudsters, in addition, use other styles of attacks on the SMS channel, such as number porting attacks, fake caller ID and call waiting scams, and subscriber identity module (SIM) clones.

– Though cheaper than proprietary hardware tokens, the SMS channel can represent a major and unpredictable financial burden. Depending on the location and message's total volume, authentication via SMS ranges in cost from $0.10 to $0.20 per transaction. Moreover, there is the cost of contracting a third-party service provider to manage text messages delivery, such as an SMS gateway provider or mobile network operator.

To mitigate this issue, the security community recommends that instead of relying on the compromised cellular networks and open SMS protocol, financial institutions must look at deploying industry-standard X.509 digital certificates to their users' mobile phones and tablets. This will allow banks to encrypt sensitive two-way communications and uniquely identify every enrolled user through their device, which is fundamentally transformed into a second factor for authentication. Also, mobiles can now be employed with absolute confidence to confirm a user's identity when logging into an online banking portal or mobile application or when performing sensitive transactions. Credit and debit card payments and call center interactions can also be authenticated in this way.

With today's refined mobile technology, there are more sophisticated, safer technologies that can considerably simplify the authentication process without sacrificing security or convenience.

2.7.3.1.2 Challenge–Response (C/R) Authentication

There are currently three common types of challenge–response (C/R) generators: Handheld C/R, Completely Automated Public Turing test to tell Computers and Humans Apart (CAPTCHA) (Ahn et al., 2003), and SiteKey. The following describes the operation of these methods in general terms, but many variations are possible.

HANDHELD C/Rs. A handheld C/R is a security mechanism for verifying the identity of a user or system without the need to transmit the actual password across the wire or wireless network. The server sends a *challenge* (string of alpha or numeric characters) to a client; this client then combines the string with its *response* (password), and from this a new password is generated. The new password is sent to the server; if the server can generate the same password from the challenge it sent the client and the client's password, then the client must be authentic.

Figure 2.13   CAPTCHA: Telling humans and computers apart automatically (Ahn et al., 2003).

CAPTCHA: CAPTCHA is a type of authentication employed to tell humans apart from machines by avoiding automated responses. It falls into the category of C/R authentication. A CAPTCHA is a program that protects websites against Internet bots (i.e. a computer program which performs automated tasks) by generating and grading tests that humans can pass but existing computer programs cannot. For example, humans can read distorted text as shown in Figure 2.13, but existing computer programs can't. It is usually a word or set of numbers and letters presented to the user that are in fact obscured or modified in some manner to prevent computers from responding to a prompt.

SITEKEY. SiteKey was a Web-based security system that offered one type of mutual authentication between end users and websites. Its main purpose was to deter phishing. SiteKey (Figure 2.14) had been employed by Bank of America (Bank of America, 2009), a large American financial institution in the United States. SiteKey was owned by RSA Security (2010). Although SiteKey has been terminated by the bank (today they use Touch ID and fingerprint), we considered this to be an interesting authentication technique for learning purposes.

SiteKey uses the following C/R techniques:

1. User identifies (not authenticates) herself to the site by entering her username (but not her password). If the username is a valid one, the site proceeds (Image **A**).
2. Site authenticates itself to the user by displaying an image and accompanying phrase that she has earlier configured. If the user does not recognize them as her own, she is to assume the site is a phishing site and immediately abandon it. If she does recognize them, she may consider the site authentic and proceed (Image **B**).

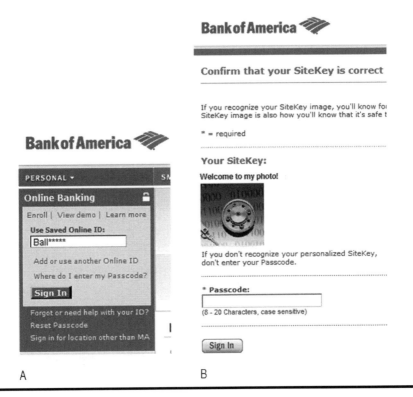

Figure 2.14 **SiteKey helps prevent unauthorized access to users' accounts. It employs a combination of username (A), then phrase, image, and passcode (B). (From Bank of America, November 19, 2007, http://www.bankofamerica.com/ onlinebanking/index.cfm?template=site_key&state=CA.)**

3. User authenticates herself to the site by entering her password. If the password is not valid for that username, the whole process begins again. If it is valid, the user is considered authenticated and logged in (Image **B**).

2.7.3.1.3 Mobile-Phone-As-A-Token Authentication

Using passwords for authentication is no longer appropriate and stronger authentication methods are required. Two-factor authentication nowadays makes use of a variety of devices such as tokens, ATM cards, mobile phones, tablets, etc. Authentication methods that leverage users' mobile phones as authentication tokens can be a viable alternative to legacy one-time password hardware tokens.

To authenticate users through a mobile phone, certain components and architecture are required, as shown in Figure 2.15. The user must have access to a computer connected to the Internet and be in possession of a mobile phone with a working SIM card. A SIM is an integrated circuit that is intended to securely store

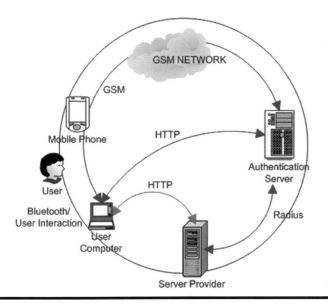

Figure 2.15   A general architecture of mobile authentication.

the International Mobile Subscriber Identity (IMSI) number and its related key, which are used to identify and authenticate subscribers on mobile telephony devices (such as mobile phones and computers).

Through the Internet browser on the computer, the user can access Web services provided by service providers (SPs). The SP is connected to an authentication server that will handle the authentication on behalf of the SP. The AS is connected to the GSM network, which enables it to communicate with the user's mobile phone and the operator's authentication center. The AS is composed of two parts, an authenticator and an Authentication, Authorization, and Accounting (AAA) server. The authenticator communicates with the client and relays messages to the AAA server, which handles the authentication. When designing an authentication scheme that uses two separate devices that communicate over two different networks, it is very important to ensure that it is the same user that controls both devices. This is called a "closed loop" going through all the components involved in the authentication, as illustrated in Figure 2.15. The loop starts in the device requesting the service (the user's computer), goes through the network with SP and AS and then via mobile phone and back to the initial device, either by user interaction or Bluetooth.

2.7.3.2 Connected Tokens

Connected tokens have to be physically connected to the computer with which the user is authenticating. It automatically transmits the authentication information to the client computer once a physical connection is made, reducing the need for the user to enter the authentication information by hand. However, in order to use a

Figure 2.16   Magnetic stripe token.

Figure 2.17   RSA SecurID® 800 with OTP.

Figure 2.18   VeriSign secure storage token.

connected token, the adequate input device has to be installed. The most common types of physical tokens are smart cards and USB tokens, which require a smart card reader and a USB port respectively.

To transfer data, most tokens must make physical contact with a reader device. Examples of such devices are magnetic stripe tokens used in ATMs* (Figure 2.16) and hardware authentication USB tokens, which are small hardware devices that can be plugged into a USB port on a computer and can also be used with a password or PIN. Examples of USB tokens are the RSA SecurID®800 Hardware Authenticator (Figure 2.17) and the VeriSign Secure Storage Token† (Figure 2.18).

---

* Retrieved on June 25, 2008 http://www.1stsource.com/personal_banking/products/resource_plus.htm.

† http://www.verisign.com/static/DEV016111.pdf.

But what if you lose your token, or lock them in your car, or what if it is stolen? This is not an uncommon problem. Losing the physical element of your two-factor authentication (2FA) will be difficult, just like losing a key for a physical lock would be. However, most services using 2FA will have procedures in place to help you work around the situation. Check your two-factor services and make sure you've done everything you need to do to prepare for the probable loss of your physical token.

2.7.4 Digest Access Authentication

Digest access authentication* (DAA) is one of the established methods a Web page can employ to negotiate credentials with a Web user. It employs the Hypertext Transfer Protocol (HTTP)[†]. This method builds upon (and makes obsolete) the basic authentication scheme, enabling a user's identity to be established without having to send a password in plaintext over the network. A DAA scheme provides no encryption of message content. The goal is simply to create an access authentication method that prevents the most serious flaws of basic authentication. One advantage of the basic authentication scheme is that it is supported by roughly all popular Web browsers.

The classical transaction is comprised of the following steps:

- The user asks for a page that requires authentication but does not provide his credentials (i.e. username and password). Typically, this is because the user simply enters the address or follows a link to the page.
- The server responds with a "401" response code, providing the authentication realm and a randomly generated, single-use value called nonce[‡]. At this moment, the system will show the authentication realm (normally a description of the computer or system being accessed) to the user and prompt for her credentials through the login window (Figure 2.19).
- Once the credentials have been provided, the system re-sends the same request but adds an authentication header that contains the response code. In this example, the server accepts the authentication and the page is returned to the user. If the credentials are invalid, the server may return the "401" response code and the system would prompt the user again.

---

* http://www.ietf.org/rfc/rfc2617.txt

[†] The basic authentication scheme was originally defined by RFC 1945 (Hypertext Transfer Protocol – HTTP/1.0).

[‡] Nonce is a random or non-repeating parameter value that is included in data exchanged by a protocol, usually for the purpose of guaranteeing liveness and thus detecting and protecting against replay attacks. A nonce can be a time stamp intended to limit or prevent the unauthorized replay or reproduction of a file.

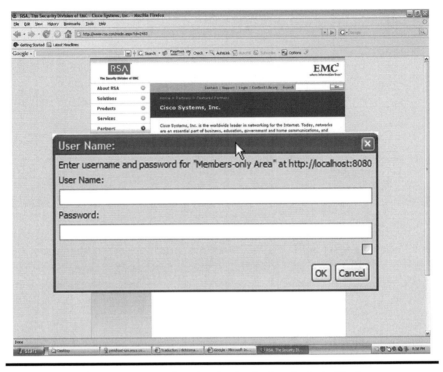

Figure 2.19   Digest Access Authentication's window login in Firefox.

2.7.5 Out-of-Band Authentication (OOBA)

Out-of-band authentication (OOBA) is the utilization of two separate networks working concurrently to authenticate a user. Therefore, it prevents an intruder from having a chance to hack the password. It is a strong defense against man-in-the-middle attacks and sophisticated online hackers. OOBA works well because even if an intruder gains all security credentials to a user's account, a transaction cannot be completed without access to the second authentication network. Consider the mobile phone scenario to verify the identity of the user involved in a Web transaction. Some of the advantages of using a mobile phone-based OOBA are the following:

- No supplementary hardware, software, or training is necessary for the end user. However, it is still difficult for the average consumer computer user to cope with.
- Users already carry phones and keep close track of them.
- Mobile phone communication can take place in true real time.
- Mobile phone authentication can require interaction with a human being.
- The public switched telephone network (PSTN) is a secure network.
- A strong, humanly understandable audit trail of the transaction is captured.

Figure 2.20   SafePass Mobile.

OOBA using a mobile phone enables legitimate account owners to be made aware of attempts to breach their accounts. If an account is protected by mobile-phone-based OOBA, the user will receive a call to authenticate a large transaction before it can be completed. If the rightful account owner is not involved in the Web transaction, he cannot complete the mobile-phone-based authentication, and the counterfeit transaction will be cancelled before losses are incurred.

To improve online security, Bank of America introduced SafePass* (Figure 2.20). To register for the SafePass OOBA service, customers add their mobile phone numbers to the accounts overview section on Bank of America's website. When a customer initiates certain online transactions, the user will be prompted to enter a six-digit code, which is sent via text message to the user's mobile phone. The code is required for transactions such as money transfers for amounts greater than the current limits, adding new bill payees, or adding new accounts for online transfers. The code expires within 10 minutes of being issued or immediately after it is used.

---

* SafePass: Link retrieved on September 23, 2015 https://www.bankofamerica.com/privacy/
  online-mobile-banking-privacy/safepass.go.

2.7.6 Risk-Based Authentication (RBA)

Similar to layered authentication, RBA requires various levels of proofs depending on the risk level of the transaction. This term is used interchangeably for systems where risk assessment is used in two different ways:

■ In some systems, risk assessment is employed to determine the toughness of the processes and procedures to sign up and use a particular set of resources. The same credentials will be employed in every session, but users who need different resources may use different credentials. A username and password will be enough for some users, whereas others, who have more access to sensitive information, may need, for example, a two-factor hardware token.

■ RBA is employed when systems require different authentication levels for the same user based on a specific transaction, not identity. For example, many Web services will use a cookie which is placed on the browser from an earlier session as proof of identity for browsing catalogue pages but will ask for a username and password to make a purchase.

An example of RBA is the RSA® Adaptive Authentication, which is an RBA and fraud detection platform that measures over 100 risk indicators to identify high-risk and suspicious activities. Adaptive Authentication is powered by RSA® Risk Engine (Figure 2.21), which conducts a risk assessment of all users behind the scenes. A unique risk score is assigned to each activity, and users are *challenged* when an activity is identified as high-risk and/or an organizational policy is violated. Adaptive Authentication monitors and authenticates activities based on risk, profiles, and policies by correlating device identification profiles, behavioral patterning profiles, user profiles, RSA® eFraud Network™ feeds, and fraud intelligence.

The RSA Risk Engine is a self-learning technology that evaluates each online activity in real-time, tracking over 100 indicators in order to detect fraudulent activity. A unique risk score, between 0 and 1000, is generated for each activity. The higher the risks score, the greater the likelihood that an activity is fraudulent.

2.7.7 Public Key Authentication

In conventional cryptography, the sender and receiver of a message know and use the same secret key; the sender uses the secret key to encrypt the message, and the receiver uses the same secret key to decrypt the message. This method is known as secret key or symmetric cryptography. The main challenge is getting the sender and receiver to agree on the secret key without anyone else finding out. If they are in separate physical locations, they must trust a courier, a phone system, or some other transmission medium to prevent the disclosure of the secret key. Anyone who overhears or intercepts the key in transit can later

Figure 2.21 **The RSA® Engine measures a number of factors in generating a risk score. (From Adaptive Authentication–Advanced Fraud Detection for Web & Mobile, 2017, https://www.rsa.com/content/dam/en/data-sheet/rsaadaptive-authentication.pdf.)**

read, modify, and forge all messages encrypted or authenticated using that key. The generation, transmission, and storage of keys is called *key management;* all cryptosystems must deal with key management issues. Because all keys in a secret-key cryptosystem must remain secret, secret-key cryptography often has difficulty providing secure key management, especially in open systems with a large number of users.

In order to solve the key management problem, Diffie & Hellman (1976) introduced the concept of public-key cryptography. Public-key cryptosystems have two primary uses, encryption and digital signatures. In their system, each person gets a pair of keys, one called the public key and the other called the private key. The public key is published, while the private key is kept secret. The need for the sender and receiver to share secret information is eliminated; all communications involve only public keys, and no private key is ever transmitted or shared. In this system, it is no longer necessary to trust the security of some means of communications. The only requirement is that public keys be associated with their users in a trusted (authenticated) manner (e.g. trusted directory). Anyone can send a confidential message by just using public information, but the message can only be decrypted with a private key, which is in the sole possession of the intended recipient. Furthermore, public-key cryptography can be used not only for privacy (encryption) but also for authentication (digital signatures) and various other techniques.

In a public-key cryptosystem, the private key is always linked mathematically to the public key. Therefore, it is always possible to attack a public-key system by deriving the private key from the public key. Typically, the defense against this is to make the problem of deriving the private key from the public key as difficult as possible. For instance, some public-key cryptosystems are designed so that deriving the private key from the public key requires the attacker to factor a large number; in this case, it is computationally infeasible to perform the derivation. This is the idea behind the RSA public-key cryptosystem*.

Another important component often used in public key authentication is a smart card. Smart cards are plastic cards that include integrated circuit cards. They are tamperproof and can be employed to store users' certificates and private keys. Smart cards can execute complex public key cryptography operations, for instance, digital signing and key exchange. It is possible to deploy smart cards (and smart card readers) to offer stronger user authentication and non-repudiation for a variety of security solutions, including logging on over a network using fingerprint (i.e. the smart card includes cryptographic keys and biometric fingerprint data), secure email, and other methods. But what are the benefits of making use of smart cards? The benefits are the following:

- Private keys are stored on the smart card (tamper-resistant) instead of, for instance, on users' hard disk (not secure). Therefore, smart cards provide stronger security for user authentication and non-repudiation.
- As cryptographic operations are disassociated from the OS, smart cards are not subject to attacks on the OS (e.g. memory dump attacks which may expose private keys or other cryptographic secrets).
- Logon credentials follow users, so the system administrator, for example, can issue a single smart card to each network user to provide a set of logon credentials for logging on to local and remote networks, which can reduce the cost of managing separate user accounts for logging on to a network and logging on remotely (Microsoft, 2009)

Additionally, because the administrative support that is needed to administer user passwords is an important cost for large organizations, smart cards can be deployed to reduce the cost of forgotten or expired passwords. Smart cards use PINs instead of passwords. The PIN protects the smart card in case of misuse due to the fact that the PIN is known only to the smart card's owner. A smart card scenario

---

* RSA cryptosystem, September 24, 2015 https://en.wikipedia.org/wiki/RSA_(cryptosystem).

is a user who inserts the card in a smart card reader that is attached to a computer and, when prompted, enters the PIN. The smart card can be employed only by a user who possesses the smart card and has knowledge of the PIN. Two concepts frequently referred to in conjunction with PKI include:

2.7.7.1 Encryption

When Alice wishes to send a secret message to Bob, she looks up Bob's public key in a directory, uses it to encrypt the message and sends it off. Bob then uses his private key to decrypt the message and read it. No one listening in can decrypt the message. Anyone can send an encrypted message to Bob, but only Bob can read it (because only Bob knows his private key).

2.7.7.2 Digital Signatures

To sign a message, Alice does a computation involving both her private key and the message itself. The output is called a digital signature and is attached to the message. To verify the signature, Bob does a computation involving the message, the purported signature, and Alice's public key. If the result is correct according to a simple, prescribed mathematical relation, the signature is verified to be genuine; otherwise, the signature is fraudulent or the message may have been altered.

2.7.8 Single Sign-On (SSO)

External compliance requirements and internal security initiatives are driving the need for more complex passwords with more frequent expirations. Without a solution that helps end users, corporate password policies only frustrate end users. Users will write down passwords or use the same password for multiple applications – weakening security and hindering regulatory compliance efforts. Additionally, supporting multiple passwords also hits the IT bottom line as employees use costly help desk resources for password resets (RSA Security, 2010).

Single sign-on (SSO) describes the ability to use one set of credentials, an ID, and a password or passcode, for example, to authenticate and access information across a system, an application, or even organizational boundaries. It may be called Web SSO when everything is accessed through a browser. With SSO, users authenticate only one time for a particular working session, in spite of where the information that they want to access is located. This process provides improved security over a simple synchronization of passwords. SSO is quite useful these days when users need to access an ever-increasing numbers of applications. SSO has major security and user benefits, as well as the capacity to reduce the help desk costs of password management significantly. There is a security risk with static password-based SSO because a breach of password security means all systems accessible by a particular user can be compromised.

A variant of the SSO is OpenID*, a decentralized subset of SSO for the Web that is different in that end users own an identity URL instead of a password. You may choose to associate information with your OpenID that can be shared with the websites you visit, such as a name or email address. With OpenID, you control how much of that information is shared with the websites you visit; your password is only given to your identity provider, and that provider then confirms your identity to the websites you visit. Other than your provider, no website ever sees your password, so you don't need to worry about an unscrupulous or insecure website compromising your identity. Also, you are able to make use of a single, existing account to sign in to numerous websites without needing to create another username and password. OpenID is an easy method of joining new sites. Yet there are also some challenges associated with using OpenID, such as the struggle of creating a username and password; however, this burden is minimized since you create your credentials only once. Still, users might not understand or may be confused by OpenID since it is not the standard username and password authentication method well known even to novice users. You should not use OpenID for high-privacy sites such as online banking, e-commerce sites, or healthcare sites since you trust only one provider for your credentials, but it is fine to use it for trivial things (e.g. library). Finally, an OpenID provider may track your habits since they receive all of the authentication requests.

2.7.9 Biometrics

Biometric authentication is being viewed to a greater extent as a credible option to knowledge- or key-based security systems. It is increasingly getting acceptance as a technology for user authentication to computerized systems. An example is the Touch ID from Apple Inc., a fingerprint recognition feature released and currently available on certain versions of iPhone and iPad. Touch ID allows users to unlock their device, as well as make purchases in the Apple digital media stores, and to authenticate Apple Pay online or in apps. Biometrics is defined as automatically recognizing a person using distinguishing traits. Biometrics techniques can be mostly divided into classical and soft biometrics. Whereas classical biometrics deals with *physical characteristics* which are unique to a specific subject (e.g. fingerprints, retina/iris, face, hand/palm, voice, odor, ear) and *behavioral characteristics* (e.g. keystroke, signature, gait, etc.), soft biometric traits refer to physical and behavioral traits, such as gender, height and weight, which are not unique to a specific subject but are useful for the identification, verification, and description of human subjects. Soft biometric attributes can be combined with classical biometric traits to improve the accuracy of a biometric recognition system. They can also be used as filters to restrict the search during a biometric identification operation.

---

* The OpenID Foundation (OIDF) promotes, protects and nurtures the OpenID community and technologies. http://openid.net/.

The extraction of soft biometric attributes can also be useful in other application domains, such as healthcare, human–computer interaction (HCI), robotics, and gaming. Classical biometrics is flexible and highly accurate. But it is very hard to get classical biometric data from a distance and without personal cooperation. Soft biometrics, although featuring less accuracy, can be used much more without restraint. Biometric authentication is typically the most powerful single-factor user authentication method. However, it is not flawless, and certain vulnerabilities lead to important security breaches given that existing biometric methods cannot provide strong authentication on their own; they require one or more authentication methods, biometric or not, in order to offer a higher level of security (i.e. strong authentication). Strong authentication combines at least two authentication factors of different types to enhance the security of the verification of the identity. Using two factors of different types as opposed to one delivers a higher level of authentication assurance. The most common example is using your debit card and PIN code to access your checking account through an ATM: the debit card is "something you *have*" and your PIN code is "something you *know*" (i.e. two-factor authentication). Another example is requesting multiple answers to security questions, which may be considered strong authentication but, unless the process also retrieves "something you *have*" or "something you *are*", it would not be considered strong (multifactor) authentication.

Biometric-based authentication applications encompass workstation, network, and domain access, single sign-on, application logon, data protection, remote access to resources, transaction security, and Web security. A biometric system operates in two modes: *enrolment and verification*. In the *enrollment* process, the user's biological characteristics are acquired and stored (*template*) for later use. This *template* is then placed in a back end database for later retrieval. In the *verification* process, the user's characteristics are measured and compared against the stored *template*. Biometric authentication is potentially the strongest single authentication method. Nevertheless, it is not infallible, and certain vulnerabilities have far-reaching consequences. Fingerprint, voice, and face recognition are the most appropriate biometrics for restricting access to Web pages, as the sensors for these authentication methods (AMs) are small, cheap, and widely available as standard options on smartphones or laptops from many computer vendors.

2.7.9.1 How Does the Biometric Authentication Process Work?

1. The data collection process captures the individual's biometric data ("Capture") using a physical biometric reader (e.g. a fingerprint scanner) and sends the unprocessed biometric data to the signal processing ("Process" and "Store");
2. Signal processing refines the unprocessed biometric data and sends the biometric sample to the matching process ("Compare").

3. The matching process ("Compare") compares the biometric sample with the individual's template and sends a score to the decision process.
4. The decision process determines whether the score is above or below some pre-established thresholds, and sends it the final Yes or No to the Application (Figure 2.22).

Physiological traits such as fingerprint or hand recognition generally require a single data scanning of the acquisition device (e.g. fingerprint scanners, iris recognition cameras, etc.) and are stable physical characteristics. On the other hand, behavioral characteristics like voice or signature recognition are more susceptible to alterations (e.g. illness, aging, emotion, etc.) and usually require multiple samples from the user in order to generate an accurate template. Therefore, behavioral characteristics require extensive user collaboration and fast data acquisition devices. The former demands ease of use of the system, while the latter at present doesn't offer data speed or user convenience since both enrollment and verification generally bother users. In theory, any good security policy means using multifactor authentication to make it more difficult for an attacker, but this undermines usability.

Two measurements often quoted as identifying the capabilities of biometric systems are the false acceptance rate (FAR) and the false rejection rate (FRR). These rates refer to the number of false negative or false positive matches returned during a biometric evaluation and verification. A FAR of a non-legitimate user may cause damage to the system it is supposed to protect. Frequent FRR of a legitimate user, on the other hand, would have undesired consequences for the user and lower user acceptance of the system. The risk of having an FRR is clear, for instance, in the case of a user having a terrible cold. The *trade-off* between the FRR and the FAR

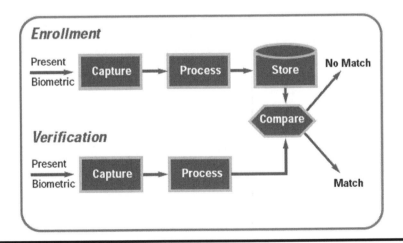

Figure 2.22 Biometric authentication process (Woodward et al., 2003).

depends on the extent that the system is able to tolerate (i.e. if the decision threshold is increased, FRR increases and FAR decreases, and vice versa) at the moment of a signature or hand verification, for example.

For a biometric system to be useful and trustworthy, it should meet specific standard requirements and characteristics. What biological measurements qualify as biometrics? Any human physiological or behavioral trait can serve as a biometric characteristic as long as it satisfies the following requirements (Radha & Kavitha, 2012):

- *Universality* (availability) states that each person should have the characteristic. It is measured by the "failure to enroll" rate.
- *Distinctiveness* declares that any two people should sufficiently have different characteristics, and it is measured by the False Match Rate (FMR), also known as "Type (II) error".
- *Permanence* (robustness) states that the characteristic should be stable (with respect to the matching features) over a period of time (i.e. stability over age). It is measured by the false non-match rate (FNMR), also known as "Type (I) error".
- *Collectability* (accessibility) affirms that the characteristic can be measured quantitatively and is easy to image using electronic. It can be quantified by the "throughput rate" of the system.
- *Performance* means to achieve recognition accuracy, speed, and the resources required for the application.
- *Acceptability* relates to the particular user population and the public, in general, should have no (strong) objections to the measuring/collection of the biometric characteristic. Acceptability is measured by polling the device users.
- Resistance to *Circumvention* tests and proves how the system resists fraudulent methods easily. Table 2.2 shows a brief comparison of the iris and retina biometric techniques based on the aforementioned factors.

Which biometric characteristic is best? Each biometric feature has its own strengths and weaknesses and the choice typically depends on the application. Accordingly, each one could be used in authentication and/or identification applications. Predicting the FARs and FRRs, system throughput, user acceptance, and cost savings for operational systems from test data is a surprisingly difficult task.

2.7.9.2 Unimodal and Multimodal Biometrics Systems

In many real-world applications, unimodal biometric systems (e.g. a single biometric method such as face recognition) often are confronted by significant limitations due to sensitivity to noise (e.g. variations in light in face sensed), intraclass variability (e.g. when a user interacts with the sensor incorrectly such as incorrect

Table 2.2 Comparison of Optical Biometric Characteristics (Gad et al., 2015)

Biometric Characteristic	Universality	Distinctiveness	Permanence	Collectability	Performance	Acceptability	Circumvention
Retina	H	H	M	L	H	L	L
Iris	H	H	H	M	H	L	L

Legend: H = High; M = Medium; L = Low.

facial pose), non-universality (e.g. some people may have scars or cuts in fingerprints, and, as a result, a fingerprint system, may extract incorrect feature details from them), and other factors. Attempting to improve the performance of individual matches in such situations may not prove to be highly effective. Multibiometric systems (Gad et al., 2015) in turn seek to alleviate some of these problems by providing multiple pieces of evidence of the same identity. Therefore, they provide more accuracy when compared to a unimodal biometric system (Radha and Kavitha, 2012). The main goal of a multimodal biometric system is to develop the security system for the areas that require a high level of security. It combines different biometric technologies (e.g. fingerprint and face recognition, or lip movement and face recognition, or iris and retina recognition, etc.) which can be very expensive to implement.

2.7.9.3 Fingerprint Recognition

Fingerprint recognition (FR) is the most widely known and commonly used biometric identification scheme. It compares users' fingerprints to a previously stored template and determines validity and authenticity based on this comparison. For example, users place their finger on a device that reads the thumbprint (Figure 2.23) on a laptop. To authenticate them, the system compares the current fingerprint pattern with a previously collected thumbprint, and access is granted only when the patterns are identical. The Apple Touch ID lets users unlock their phones by using

Figure 2.23   Apple iPhone Touch ID. (https://support.apple.com/en-ca/HT201371 accessed 08/24/2018.)

their fingerprints. It uses sophisticated algorithms to recognize and securely match their fingerprints.

Fingerprint recognition technology is divided into two distinct processes: *verification* and *identification* (Figure 2.24):

■ In the *verification* process, the user states who he/she is, and a fingerprint is taken and compared to the user's previously registered fingerprint. If the fingerprint matches, the user is "verified" as who he/she says he/she is. Since the newly acquired fingerprint is compared to only one stored fingerprint, this is called a one-to-one matching process (1:1). As in the enrollment process, when fingerprint verification is done, only the fingerprint template is used in the comparison, not the actual image of the fingerprint.

■ In the *identification* process, the user doesn't need to state who he/she is. A fingerprint is taken and compared to each fingerprint in the database of registered users. When a match occurs, the user is "identified" as the existing user the system found. Since the newly acquired fingerprint is compared to many stored fingerprints, this is called a one-to-many matching process (1:N). As in the verification process, when the fingerprint verification is done, only the fingerprint template is used in the comparison, not the actual image of the fingerprint.

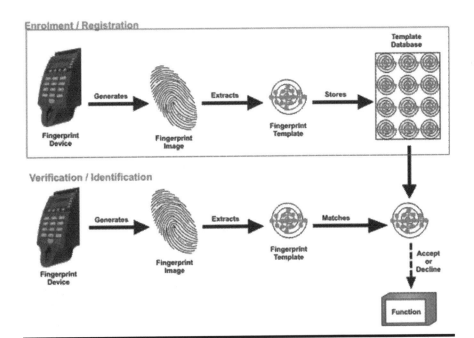

Figure 2.24 Fingerprint recognition scheme.

2.7.9.4 Optical Recognition

To tie identity more closely to an individual and have a more appropriate authorization, the use of optical methods for biometric authentication is increasingly gaining user acceptance. Currently, the most widespread types of optical authentication methods are the retina and iris recognition. Retina recognition (RR) aims to identify a person by comparing images of the blood vessels in the back of the eyes, the choroidal vasculature. On the other hand, iris recognition (IR) takes an infrared picture of the iris, the colored part of the eyes. This is considered by users to be less intrusive. These two types of biometrics have come a long way in recent years, allowing the individual to be scanned even through their glasses or contact lenses. Iris authentication systems are commercialized by Panasonic* and Iris ID Systems[†], and retinal systems by Zoloz (Eyeprint)[‡]. RR and IR authentication methods may be employed alone or integrated with other technologies such as smart cards, encryption keys and digital signatures (Gad et al., 2015), (Hoanca and Mock, 2006).

2.7.9.5 Facial Recognition

An image is examined for overall facial structure. In authentication applications, the system has a camera that searches a user's face and matches it against the face stored in the user record. A popular face recognition solution is Face ID on the latest iPhone X. Face ID lets you securely unlock your phone, authenticate purchases, sign in to apps, and perform other tasks with just a glance[§].

One of the recognized commercial face recognition systems is produced by NEC, which produces the NeoFace® Smart ID as shown in Figure 2.25. NeoFace® Smart ID is a positive, multimodal biometric identification. It is a mobile application for commercial off-the-shelf smartphones and tablets. The solution enables users to capture multimodal biometric data, including facial but also fingerprint and voice. NeoFace Smart ID can also be configured for submittal to a central database or with a local facial watch list allowing for on-the-spot identification prior to remote database search (Figure 2.25).

2.7.9.6 Voice Recognition

Voice recognition is well suited for telecommunications applications, and the latest mobile devices already have the necessary hardware to utilize these applications. Voice recognition systems prompt users for some spoken words and authenticate them based on distinguishing speech patterns (voiceprints).

---

* ftp://ftp.panasonic.com/pub/Panasonic/CCTV/SpecSheets/BM-ET300.pdf.

[†] http://www.irisid.com/productssolutions/irisaccesssystem.

[‡] http://www.zoloz.com/technology.

[§] https://support.apple.com/en-ca/HT208109.

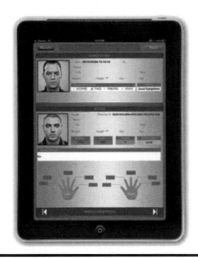

Figure 2.25   NeoSmart ID, facial recognition from NEC. (https://www.necam. com/docs/?id=d0eab181-e5a2-46ac-82cd-44e7a0e84fcd.)

One of the drawbacks of voice recognition is the impersonation attack, where an unauthorized individual changes her biometric to appear to be an authorized individual. Also, it is more susceptible to alterations (e.g. illness, aging, emotion, etc.).

Nuance is one of the vendors which produce voice recognition systems for user authentication currently offering two types of voice biometrics such as VocalPassword (authentication by passphrase) and FreeSpeech (authentication by conversation)*. It makes use of a customer's unique voiceprint for authentication. It can be passive, where the user can say anything and their voice is matched to a voiceprint. Or it can be active, where the caller is asked to recite a passphrase, as shown in the Figure 2.26.

2.7.9.7 Signature Recognition

Signature recognition is a visual/behavioral biometric system. It operates in a three-dimensional environment. It measures height and width as well as the amount of pressure applied in a pen stroke. Dynamic signature verification (Figure 2.27) takes into account how the signature was produced. It is the alterations in speed, pressure, and timing that take place during the act of signing that are relevant, and not the shape or look of the signature. In fact, only the original signer can reconstruct those alterations in timing and X, Y, and Z pressure.

Signature authentication software is produced by Cyber-SIGN. There are two key types of digital handwritten signature authentication, static and dynamic. Static

---

* https://www.nuance.com/omni-channel-customer-engagement/security/multi-modal-biometrics/freespeech.html.

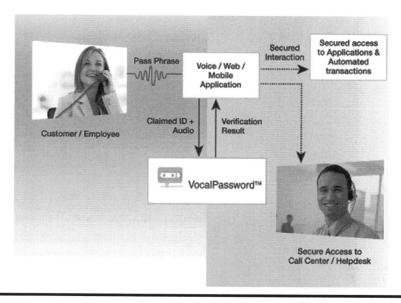

Figure 2.26 VocalPassword™ (authentication by passphrase) by Nuance. (https:// www.nuance.com/omni-channel-customer-engagement/security/multi-modal-biometrics/vocalpassword.html.)

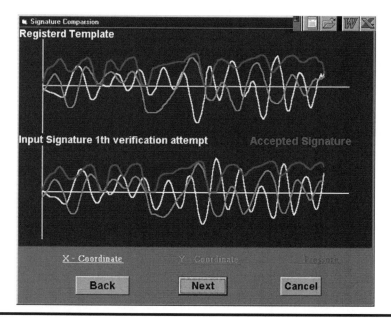

Figure 2.27 Dynamic signature verification from Cyber-SIGN. (http://www. planetpdf.com/forumarchive/Cyber-SIGN%20Acrobat%20eu(2).pdf.)

is most often a visual comparison between one scanned signature and another scanned signature, or a scanned signature against an ink signature. Technology is available to check two scanned signatures using advanced algorithms. Dynamic signature authentication is becoming more popular as ceremony data (i.e., the action of signing initiates a "signing ceremony") is captured along with the X, Y, T, and P coordinates of the signer from the signing device. This data can be utilized in a court of law using digital forensic examination tools, to create a biometric template from which dynamic signatures can be authenticated either at time of signing or post signing, and as triggers in workflow processes.

One example is the Cyber-SIGN Acrobat, which is a plug-in for Adobe Acrobat that provides the ability to electronically "sign" a PDF document with handwritten signatures. Using Cyber-SIGN and built-in Acrobat Digital Signature Framework, it gives the users the ability to embed legally compliant electronic signatures into the documents.

2.7.9.8 Keystroke Recognition

In this type of technology, not only must the attackers know the correct password, but they must be also able to reproduce the user's rate of typing and intervals between letters. Plurilock's* BioTracker uses behavioral biometrics – mouse movements and keystroke rhythms – to authenticate users as they log in with a username and password. The software then continues to analyze these behavioral biometrics during the user's session to protect against situations in which a legitimate user logs on but the session is then continued by an unauthorized user.

As illustrated in Figure 2.28, BioTracker is implemented as client/server software, with two different types of clients: application-based and Web-based. Application-based clients can be used to monitor individual workstations or virtual machines (in the cloud), while Web-based clients are plug-ins used to monitor websites. The client module is responsible for mouse movement and keystroke data collection. This data is sent to the server software, which is in charge of analyzing the behavioral biometrics and computing a biometric profile. The computed profile is then submitted to a behavior comparison unit, which checks it against the stored profiles.

2.7.9.9 Advanced User Authentication Methods

This section provides two examples of advanced and novel user authentication methods: GlanceID, an optical authentication system, and AuthenLink, an under-skin RFID chip user authentication system.

---

* Keystroke Dynamics: https://www.plurilock.com/lp/msp/.

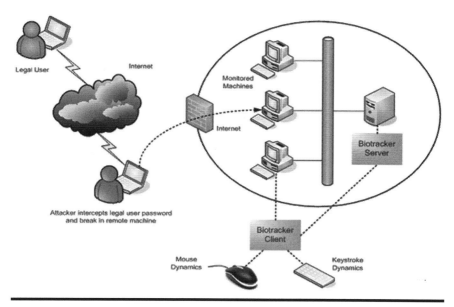

Figure 2.28 Keystroke dynamics.

2.7.9.9.1 GlanceID Optical Authentication

An example of a unimodal biometric authentication method providing the same level of security as a multimodal system using a single optical authentication is GlanceID (Braz & Seffah, 2016), a glance-based and "two-factor" optical authentication method with strong authentication capability.

As already mentioned, the goal of a multimodal biometric system is to develop the security system for the areas that require a high level of security. It combines different biometric technologies (e.g. fingerprint and face recognition, or lip movement and face recognition, or iris and retina recognition, etc.) which can be very expensive to implement. To provide this level of security using a single optical authentication, we propose GlanceID. This method comprises of something you *are* (eye gaze) and something you *have* (objects positioning) in a single biometric authentication method. It captures and processes the eye gaze direction and objects positioning in real time with the aim of granting access to users for physical access control. Moreover, if we consider that "objects" and "positioning" are two different entities (which they are), we can claim that GlanceID is a multi-factor authentication: something you *are* (eye gaze), something you *have* (objects), and something you *know* (positioning).

2.7.9.9.1.1 "Slippy" User Experience – The Glance-Ability Factor   "Slippy" user experience (UX) is about glance-ability and focuses less on engagement but more on quick task-based user requirements. It describes a natural evolution from sticky user experiences, for which the design goal is to get a user to notice, then

stick around and keep using your website, application, or product. On the contrary, the goal when designing a slippy UX is for it to catch a user's eye, then seamlessly integrate with that person's life and support whatever he or she needs. This type of experience lets people get on with their life while it does useful things for them. Think, for example, of user interfaces for an airplane, a nuclear reactor, or a military application. This type of experience should *enhance, not interfere* with what a user is doing; nor should it consume too much of a user's attention. For example, an Apple Watch by Apple will often be a task-based operation; it would particularly require a slippy UX. Additionally, the lack of real estate means that *glance-ability and scannability* become crucial. With its health and fitness integration, the Apple Watch is a good example of how just "glancing" at the screen can provide you with all the information you need in a snapshot. Another relevant example is the GlanceID, which is the main topic of this paper and which is described in the next paragraphs.

2.7.9.9.1.2 GlanceID Recognition   GlanceID is an optical authentication method with a strong authentication capability that uses *eye gaze direction* and *object positioning* in real time in order to grant access to users for physical access control. As mentioned, this represents a two-factor authentication such as something you *are* (eye gaze) and something you *have* (object and positioning) in a single biometric authentication method. With a wide field of view provided by GlanceID, users are free to demonstrate natural human behavior (i.e. eye gaze), and not to be forced to act in certain ways with the purpose to authenticate themselves to a system (e.g. Google Authenticator: users first need to install the authenticator app on their smartphone; to log into a site or service that uses two-factor authentication, they provide username and password to the site and run the authenticator app, which produces an additional six-digit one-time password; then users provide this to the site, the site checks it for correctness and authenticates users). The *object positioning* can be realized by the precision tracking mode, with the screen calibrated. Within one second and a half, a user is reliably authenticated, providing a quite secure and easy-to-use optical authentication method. It is especially recommended for physical access entry for high-security facilities such as military installations, nuclear facilities, laboratories, research centers, and so on. For example, military and security organizations are more likely than civilian organizations to have more stringent security needs. Our system works with all eye types, with or without contact lenses and with most eyeglasses from a distance up to one meter without compromising system accuracy. It works in bright sunlight or at night time. We illustrate the physical access control scenario in the next section.

2.7.9.9.1.3 How Does GlanceID Work?   In the enrollment process, a "sight point" (and its location) is chosen by the user at a specified and authorized door perimeter. It might be a light switch or a decorative iron details on the wall next to the door, or it can be anything that is successfully identifiable by the user. The sight point is her credential which is presented to the system through the eye gaze direction reading at the moment of the verification process.

The following describes an example of the GlanceID system in a physical access control scenario (Figure 2.29):

Bob needs to authenticate to a high-security military facility:

STEP 1: Bob approaches the operational range (e.g. a marked area on the floor) of the authentication unit (camera and sensors) next to the sliding door;

STEP 2: Proximity sensors detect Bob and trigger a small digital stereo camera pair (i.e. acquisition device) to precisely capture his gaze direction;

STEP 3: Then the proximity sensors detect his object positioning (i.e. sight point), which is then streamed over the network to the authentication server;

STEP 4: Next, the AS proceeds with the authentication process, finding Bob's sight point stored in the database;

STEP 5: The AS compares the trial template against Bob's reference template;

STEP 6: If the parameters of template correspond (Bob's gaze direction and sight point positioning), the AS successfully authenticates Bob to the system, which makes the sensors unlock the sliding door granting him access to the high-security facility.

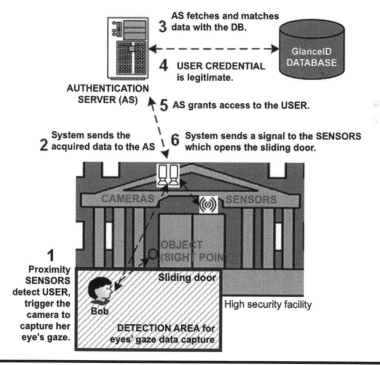

Figure 2.29   GlanceID: Optical authentication method architecture (Braz and Seffah, 2016).

GlanceID is not a multimodal authentication method since it only makes use of a single authentication mechanism. Although multimodal biometric authentication can also increase the level of security achieved by the system, GlanceID provides relatively the same high level of security without the corresponding high implementation costs while "acting" as a multimodal authentication. The GlanceID platform incorporates properly into existing security systems or operates as a standalone. Although GlanceID has not been implemented, its feasibility has been demonstrated theoretically and the results have been validated experimentally.

2.7.9.9.2 AuthenLink: Under-Skin RFID Chip Authentication

AuthenLink* is a user authentication method dealing with human-implanted RFID chip developed by (Braz & Aïmeur, 2003). It is a wireless and user-centered authentication system to authenticate humans to remote systems. It is specifically designed to protect against fraud, counterfeit, and theft, and it is particularly well suited for high-risk security systems. The system achieves its goal through a microprocessor-chip (ChipTag) computer implanted under human skin. This ChipTag is able to authenticate a user's access to systems, connect them wirelessly through the RFID technology, and enable mobile devices (smartphones, tablets, or any other mobile reader with an antenna embedded) to perform mobile transactions, access files, and so forth.

2.7.9.9.2.1 How Does AuthenLink Work?
Figure 2.30 shows the system's components.

STEP 1: When the antenna-embedded mobile reader (MR) is activated, say, when the ChipUser turns on the MR, it radiates a small amount of radio frequency energy through its antenna onto the ChipTag.

STEP 2: Radio frequency energy passes through the skin, energizing the inactive ChipTag, which then emits a radio frequency signal conveying the ChipUser's unique ID to the MR for the purposes of user authentication. Using the energy it receives from the signal when it enters the radio field, the ChipTag will briefly converse with the MR for verification and data exchange. The ChipTag has no power supply, and a tiny transmitter on the ChipTag sends out the data (unique ID).

STEP 3: Once that data is received by the MR, it automatically authenticates the ChipUser's ID with the AS by means of a base station (cell phone tower), and a Universal Mobile Telecommunications System (UMTS) mobile network through the Internet. A Secure Shell (SSH) session automatically logs the ChipUser onto a remote AS. The ChipUser gives his or her public key to the AS and then, when it connects, the AS knows access is permitted and automatically enables the connection. In fact, SSH uses a public/private encryption system to authenticate the ChipUser to the AS

---

* "A User Centred Mobile RFID-Based Authentication System for a Secure Access," master thesis, Master of Science in Electronic Commerce, University of Montreal http://christina-braz. bucket.s3-website-us-west-1.amazonaws.com/education.html.

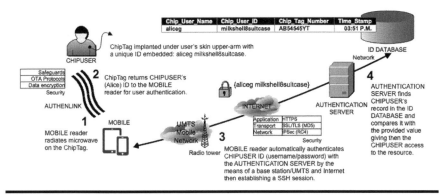

Figure 2.30 AuthenLink: UMTS architecture mode – maximum mobility.

without the intervention of the ChipUser. We merely need to create a public/private key pair for the ChipUser and then store the public key on the AS. Then, our SSH session client can use that key pair to automatically authenticate the ChipUser to the AS.

A program may also be installed on the MR and executed in the background as part of the OS. For example, Alice wants to connect with the company's Intranet website to access an important file. When Alice enters the Intranet's page, an SSH session is opened, and the AS automatically recognizes Alice's ID by retrieving information from the database about the corresponding ChipUser, thereby giving her access to the desired resource.

STEP 4: Once the data is received by the AS, it can be sent to the database for processing and management. Linking each ID from the database to the ChipUser is performed.

2.7.10 Kerberos

Kerberos* is an authentication method created by MIT – Massachusetts Institute of Technology (http://web.mit.edu/kerberos/) as a solution to network security problems. Kerberos is a network authentication protocol that supplies strong authentication and shares temporary base secrets for client/server applications by using secret-key cryptography. It uses strong cryptography so a client may prove its identity to a server (and vice versa) across an unsecure network session. Thus, all authentication processes happen between clients and servers. In Kerberos ontology, a "Kerberos client" is an entity that obtains a service "ticket" for a Kerberos service. A client is commonly a user. The designation "Kerberos server" usually appeals to the key distribution center (KDC). The KDC carries out the authentication service and the ticket granting service (TGS). The KDC has a copy of each password

---

* The name Kerberos comes from Greek mythology; it is the three-headed dog that guarded the entrance to Hades.

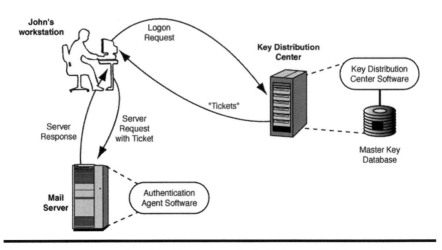

Figure 2.31   A network authentication method: Kerberos (Smith, 2002).

related to every client or server. Hence, it is crucial to keep the KDC as secure as possible.

How does Kerberos work? An indirect authentication design pattern, which appears in the deployment of an authentication system like Kerberos, works according to Figure 2.31:

- John logs on to the KDC, which supplies him with two encrypted credentials that are designated *tickets* (one encrypted for his master key and the other encrypted with the mail server's master key).
- Then, John decrypts his *ticket* to gather the shared secret key (e.g. something that the user and the system hold in common: a password) and forward the other to the email server, which uses that *ticket* to authenticate John. In fact, a *ticket* is an encrypted copy of a temporary base secret, and it is encrypted with the master key known only by the KDC and the *ticket*'s supposed receiver.

Chapter 3

Usable Security Concerns Related to Authentication Methods

This chapter reviews the user authentication methods from the perspectives of usability, security, and their potential conflict. The goal is to highlight the major usable security concerns which have been reported either in literature review or via industry case study. For each of the methods described in the previous chapter, we address the usable security concerns.

Authentication is an enabling task that needs to be completed to get to the resources required to do real work (Sasse et al., 2001). The function of user authentication should not be to educate users to better manage their security (authentication) issues, nor should it be so difficult to use that it requires either mental strain or the knowledge of a long series of rules to observe (Kerckhoffs, 1883). Instead, it should provide the least user interaction, meaning that authentication procedures are unobtrusive, involving almost no user input and are intuitive, helping users to authenticate themselves. Gartner defines "user authentication" as the real-time corroboration of a claimed identity with an implied or notional level of trust. This is a foundational identity and access management (IAM) function because without sufficient confidence in users' identities, the value of other IAM functions – such as authorization (especially segregation of duties), audit, and analytics – is eroded (Allan et al., 2014). Good examples of the latter are the "zero" interaction authentication (Corner and Noble, 2002), and, as shown in Figure 3.1, the RSA SecurID Software Token: RSA Security Toolbar with an autofill code feature that drastically reduces user interaction.

Figure 3.1 RSA Security Toolbar: AutoFill code improving usability in user authentication.

The usability of user authentication mechanisms has seldom been investigated, and since security mechanisms are conceived, implemented, put into practice, and violated by people, human factors should be taken into account in their design (Adams & Sasse, 1999). For example, social engineering attacks precisely target the human link, and they represent a very effective attack vector. A reformed and world-famous controversial computer hacker, Kevin Mitnick, found that he never had to crack passwords by technical means because he could constantly get them from people. Therefore, predicting user behavior is crucial in order to mitigate or eliminate the risk of social engineering. This is achieved in this book by using the most widely used method of cognitive task analysis called the Goals, Operators, Methods, and Selection Rules (GOMS) model (Card et al., 1983) according to Proctor and Vu (2005). The GOMS model is the general term for a family of human information processing techniques that attempts to model and predict the user behavior mentioned previously.

Now, we discuss usability aspects relevant to some of the authentication techniques discussed in the previous chapter.

3.1 Usability Concerns with Knowledge-Based Authentication (KBA)

3.1.1 Passwords

In an extremely networked world (as, for example, the Internet) wherein users may access numerous applications, password protection is regarded as expensive, awkward, and insecure. The requirement of authentication to access different applications, services, or facilities might generate frustration among users on a daily basis because users might need to frequently access the same secured applications in short periods of time.

Passwords are the first line of defense against attacks on a computer system. The rules for password choice can definitely be a burdensome problem for users and a security problem for systems. For example, trivial choices that are easy to guess are broken within seconds using password-cracking techniques – the longer the password the more difficult it is to crack. To prevent hackers from gaining access to a company's computer, experts recommend using a complex password (i.e. strong password) which in a first instance increases the short-term memory (STM) load of users, causing frequent errors. The capacity of STM is usually limited to $7 + 2$

items (e.g. letters, digits, words, etc.) (Miller, 1956). Traditional password systems include many design features for the purpose of making trial-and-error attacks as difficult as possible. Actually, they violate most of the recognized usability standards for computer systems. From the eight "Golden Rules" for interface design recommended by Shneiderman and Plaisant (2005), as shown in Table 3.1, password interactions break six rules. Table 3.2 presents the "six password broken rules" and their corresponding usability issues through a couple of examples and security benefits. Users should also follow a set of rules (i.e. password requirements in a security policy) especially related to password creation. "Your password must contain: 8 to 32 characters, at least 6 alphabetic characters (a mix of upper and lower cases), at least 1 non-alphanumeric character, not allowed: @<>&%. You may not reuse one of your last 3 passwords. Do not write anything down" (EMC, 2009). If users don't pay proper attention to this password policy, their accounts become vulnerable to intrusion by spoofers.

Users have too many passwords with different values, expiration periods, and composition rules. Usually these new complexity requirements drive users to forget their passwords and call the help desk, write down their passwords, and finally chose common, insecure passwords. Despite their well-known security and usability issues, passwords are still the most popular method of end-user authentication.

Guessing and offline dictionary attacks on user-generated passwords are often viable due to their limited entropy. According to Vance (2010), 20% of passwords are covered by a list of only 5,000 passwords. Thus, intending to increase security, complex password policies are often enforced by requiring, for example, a minimum number of characters, inclusion of non-alphanumeric symbols, or frequent password expiration. This regularly creates a detrimental conflict between security

Table 3.1 Do the 8 Golden Rules of User Interface Design Apply to Security Systems? (Shneiderman & Plaisant, 2005)

Golden Rules of User Interface Design	*Adequate for Passwords?*
1. Strive for consistency	Yes
2. Enable frequent users to use shortcuts	*No*
3. Provide informative feedback	*No*
4. Ensure Dialogues yield closure	Yes
5. Prevent errors and provide simple error handling	*No*
6. Enable easy reversal of any action	*No*
7. Put the user in charge	*No*
8. Reduce short-term memory load (F)	*No*

Table 3.2   The "Six Password Broken Rules" and their Corresponding Usability and Security Mismatches (Shneiderman & Plaisant, 2005)

Item	Usability	Security
Don't enable Frequent users cannot use shortcuts	Users can't take shortcuts: the system won't match the first few letters typed and fill in the rest	Prevents dictionary[a] and eavesdropping[b] attacks
Don't provide informative feedback	Users hardly see the password they type: they can't find out repeated letters/accidental misspellings	Prevents guessing attacks and Social Engineering[c]
Don't prevent errors and don't provide simple error handling	Most systems only mention success or failure: they don't show how close the password guess was, or even discern between a mistyped username and password	Prevents guessing, eavesdropping, and social engineering attacks
Make reversal of any action difficult	Most systems keep track of incorrect guesses and take irreparable action (locking the user's account) if several bad guesses happen	Prevents guessing, eavesdropping, and social engineering attacks
Don't put the user in charge	The system makes users be "responders" of actions rather than the initiators.	Prevents guessing, eavesdropping, and social engineering attacks
Don't reduce short-term memory load	Users must follow a set of security policies related to password creation recommended by EMC, (2009). STM is usually limited to 7+2 items	Prevents guessing, eavesdropping, and social engineering attacks

[a] A form of attack in which an attacker uses a large set of likely combinations to guess a secret.
[b] Electronic eavesdropping is the intentional surveillance of data: voice, fax, email, mobile telephones, etc. often for nefarious purposes;
[c] To infiltrate a physical building or information systems using non-technical means (e.g. searching user desks for passwords on notes).

and usability – as already highlighted in the context of password selection, management, and composition – and compels users to find the easiest password that is policy-compliant.

3.1.2 Security Questions

As discussed previously, a good security question has characteristics such as being easy to remember, even five or ten years from now; as a minimum, thousands of possible answers (not a question you would answer, for example, on Facebook in a "Fun Questions to Ask" survey or in an article or interview); having a simple one- or two-word answer; having an answer that never changes.

The following lists a couple of examples of questions and the reasons why they are good security questions from the usable security point of view:

- ■ *What was the last name of your third-grade teacher?*
  - – *Why*: Teachers change over time and most schools will have multiple teachers for each grade.
- ■ *What was the name of the boy/girl you had your second kiss with?*
  - – *Why*: First kiss seems too obvious, and it is unlikely that you have gone into great detail online about your second kiss.
- ■ *Where were you when you had your first alcoholic drink (or cigarette)?*
  - – *Why*: Unless you are a teenager and you posted online how excited you were for your first beer, it's unlikely you answered this anywhere. Use a specific location and avoid answers like home, school, or work.
- ■ *What was the name of your second dog/cat/goldfish/etc.?*
  - – *Why*: First pet's name is too obvious, but only use this question if your second pet isn't your current pet.
- ■ *Where were you when you had your first kiss?*
  - – *Why*: Even if you talked about having your first kiss online, it's unlikely you went into great detail about where you were. Just make sure the answer is short and not obvious like the name of your high school.
- ■ *When you were young, what did you want to be when you grew up?*
  - – *Why*: Only use if the answer is not cop, doctor, firefighter, or other very obvious answers.
- ■ *What is John's (or other friend/family member's) middle name?*
  - – *Why*: Since most people will not know who "John" is. You can also use their information for hard to guess security questions like "What was the name of John's first dog?"
- ■ *What is the first name of the person who has the middle name of Herbert?*
  - – *Why*: It is very unlikely you posted this anywhere, and since most people do not have their full names online this would make a great security question.

3.2 Usability Concerns with Single Sign-On

Typically, SSO is used in conjunction with some form of two-factor authentication. Although security vendors such as Centrify, Microsoft, Ping Identity, CA Technologies, and others claim the benefits of SSO* as convenient and easy to use, users still need to create a secure (strong) password, which is a cumbersome task and difficult to remember, which increases memory workload. This is the very same usability problem encountered with traditional password systems. Furthermore, it is also difficult to design an easy to use, secure, privacy-preserving, and simple-to-implement Web SSO system that could drive adoption by Web users and relying parties (RPs). These challenges entail complex trade-offs, dependencies, constraints, and sometimes even paradoxical design requirements.

3.3 Usability Concerns with CAPTCHAs

CAPTCHA codes are increasingly being used to differentiate between humans and machines (e.g., I'm not a robot), especially signing up to a website, downloading a large file, etc. From a security perspective, it is perfectly alright, as machines are mostly not able to predict the content of CAPTCHA codes and are therefore unable to create fake accounts, initiate downloads, etc. In other words, machines pretending to be humans would not be able to utilize the server resources, thus protecting the server from denial-of-service (DOS) or distributed denial-of-service (DDOS) attacks. But CAPTCHA codes can create botheration for legitimate human users. More and more complex CAPTCHA codes are being developed to protect against sophisticated machine algorithms that can detect the basic captchas, thus increasing the security and, as a by-product, severely affecting the usability (e.g., annoyingly extending the time users need to achieve their goals and making the task at hand more difficult).

The attackers today are developing more and more sophisticated algorithms that are able to detect basic CAPTCHAs. To counter this situation, more and more complex CAPTCHAs are developed to protect the systems, thus creating additional usability issues. This calls up for enhancing the usability of security to assist human users. Figure 3.2 presents some of the complex CAPTCHAS in place today which raise serious issues pertaining to the usability of CAPTCHAS for legitimate human users.

3.4 Usability Concerns with Public Key Authentication

The "Usability of Security: A Case Study" (Whitten and Tygar, 1998) was developed in order to evaluate the usability of Pretty Good Privacy (PGP) 5.0. PGP,

---

* How to Make the Right Choices for Access Management and Single Sign-On. Gartner Research, published 18 December 2017.

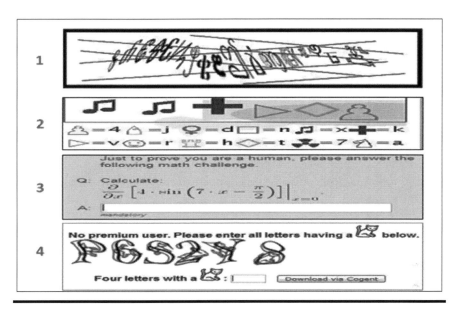

Figure 3.2   Complex Captcha Codes.

developed by Phil Zimmermann, is a software standard that uses public key infrastructure to encrypt, decrypt, and digitally sign data for the encryption of email. The study's authors chose PGP because it has a very good user interface by established standards and they wanted to know whether that was sufficient to allow non-programmers who know little about security to in fact use it effectively. The most meaningful results, obtained from a cognitive walkthrough and user test methods, unequivocally show that users had considerable difficulty with the following:

- Avoiding dangerous errors.
- Encrypting messages.
- Understanding the public key model.
- Figuring out the correct key to encrypt with and how to encrypt with any key.
- Decrypting a message.
- Publishing the public key.
- Verifying a signature on an email message.

Well, all these items listed above are in fact the basic tasks needed to run the program correctly! Therefore, PGP has not been considered usable as a way to provide effective security for most email users, according to the authors, because of the fact that there is a "mismatch between the design philosophy behind its user interface and the usability needs of a security utility".

3.5 Usability Concerns with Advanced Biometrics

3.5.1 GlanceID

The main security advantage with biometric authentication is that the credentials are truly unique to the user and are often difficult to forge. They cannot be lost, stolen, or forgotten, eliminating threats like guessing, social engineering, and brute force and point of entry attacks. However, capture and replay attacks of the digital code are possible when traveling over the network; the same kind of attack is also possible with GlanceID. Nevertheless, the main differentiation compared to the other optical authentication systems is that the object positioning (sight point) mechanism provides additional information (i.e. something the user *knows*, which is a user credential), which represents, in fact, a two-authentication factor. A security flaw of GlanceID, just as with biometrics in general, is that if someone succeeds in forgery or replay, there is no way to disable the credential and issue a new one. You cannot simply replace someone's eyes with new ones. It is also likely that many threats to optical biometrics that are not yet known will become apparent once there are more large-scale applications using biometrics and thus a greater reason for attack. Biometrics systems usually are subjected to three categories of attacks:

1. Trial-and-error attack: It is a classic way of measuring biometric strength;
2. Digital spoofing: An attacker transmits a digital pattern that mimics that of a legitimate user's biometric signature; this is equivalent to password sniffing and replay. Biometrics can't prevent such attacks by themselves (Smith, 2002);
3. Physical spoofing: It presents a biometric sensor with an image that mimics the appearance of a legitimate user.

But how does GlanceID cope with these attacks and protect the system?

With regard to attack 1: Although the attacker might be able to get the right biometric pattern, s/he cannot disclosure the sight point "secret";
With regard to attack 2: Even if the attacker transmits a digital pattern, s/he is not able to transmit the sight point "secret" over the "network" since it is intangible and cannot be copied;
With regard to attack 3: The attacker is not able to present the exact same user eye gaze direction.

Organizations may diverge in the capabilities they need for their security and usability requirements. Figure 3.3 shows a model of a potential scale of authentication factors' characteristics from lowest security to highest security and from lowest usability to highest usability. As this diagram indicates, the usability of the authentication method is likely to decrease in proportion to increasing security requirements. Additionally, interoperability requirements (e.g. storing multiple certificates

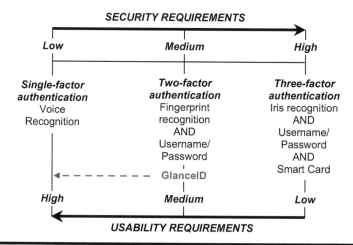

Figure 3.3 The Security and Usability Requirements scale shows the different authentication factors' characteristics from lowest security to highest security (upper arrow) as wells as from lowest usability to highest usability (lower arrow).

in a smart card) will impact the capability, size, and cost of the card. Given that all data transmission, from the camera to the object, from the object to the camera to the sensors to the authentication server, is encrypted, it is extremely difficult to undertake man-in-the-middle attacks. Another way the sight point can be revealed is when the user shares it with another person.

There is no need to create or store digital templates (i.e. pictures) of the eyes; therefore, eye gaze direction or object positioning cannot be reconstituted to generate any sort of visual image. Moreover, GlanceID can record the duration of fixation on the sight point object and serves as another layer of strong authentication. Another security advantage of GlanceID is against the well-known trial-and-error attack; it is, in fact, a cognitive safeguard and not related to commonly hardware and logical ones. Consider this attack scenario: an attacker tries to figure out the sight point secret; s/he looks at every square inch close to the door entry, which is usually the first attempt to find the object position; however, looking at every inch is tremendously tiresome from a cognitive standpoint, and even doing a small chunk of five inches by five can be considered an extremely strenuous task.

As Figure 3.3 points out, the usability of the authentication methods is likely to decrease in proportion to increasing security requirements. It is possible to combine different approaches, such as single-factor authentication to multi-factor (two or more) authentication, to achieve a higher level of security. As shown in Figure 3.3, GlanceID is positioned in the Security and Usability Requirements scale as a strong authentication method (i.e. Medium Security Requirements) even though it presents a single-factor authentication. Because of that, GlanceID presents, in fact, Low Usability Requirements. Another interesting point is that our optical system does not require re-enrollment (e.g. changes in voice tone may trigger re-enrollment process), thus providing suitable user convenience.

3.5.2 Usability Concerns with Biometrics

From the usability perspective, the ideal systems only require a single biometric signature to enrol a user. For example, fingerprints and hand palm recognition typically require a single reading, whereas behavioral systems like voice recognition or written signatures are more exposed to variation and frequently require multiple readings to train the system to recognize each user. The secret is to balance the likelihoods of false acceptance rate (FAR) and false recognition rate (FRR) so that the system rarely locks out legitimate users and it doesn't fall for masquerades.

The FAR is the measure of the likelihood that the biometric security system will incorrectly accept an access attempt by an unauthorized user. A system's FAR typically is stated as the ratio of the number of false acceptances divided by the number of identification attempts.

The FRR is the measure of the likelihood that the biometric security system will incorrectly reject an access attempt by an authorized user. A system's FRR typically is stated as the ratio of the number of false rejections divided by the number of identification attempts.

According to Smith (2002), as shown in Figure 3.4, biometric readings from a particular user should closely match that user's biometric pattern; at worst, the difference should never exceed the matching threshold shown by the dashed line. The risk of masquerades (i.e. FAR) is reduced by moving the curve to the left, which reduces the amount of gray area on the left. But this reduces usability, given that it increases the likelihood of FRRs (the gray area on the right).

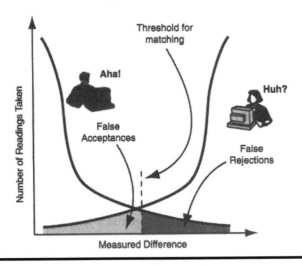

Figure 3.4 Balancing security and usability (Smith, 2002).

3.6 Comparative Analysis of User Authentication Methods

As part of one of the tasks to understand what authentication methods are, how they work, and what different features they have, a comparative analysis (based on discussion in Chapters 2 and 3) of the main user authentication methods is presented in Table 3.3.

The frame of reference for this comparative analysis has been acquired from specific sources indicated in the footnotes and also from observation, experience, and secondary data. It has been established according to the different attributes, such as characteristics/acquisition device; definition; advantages; disadvantages; security; costs; usability and place of use; acquisition time; industrial application; and, finally, accuracy, that are contained or possessed by the authentication methods. The primary grounds for comparison have to do with the most representative authentication methods currently in the market and also a novel advanced user authentication method created by the authors which has not yet been commercialized.

The following paragraphs include explanations of the definitions, abbreviations, notes, footers, and sections used in Table 3.3:

- To describe the authentication methods' attributes, subjective rating scales have been used, such as:
  - **"Security" and "Usability"** range from 1 = Minimum to 5 = Maximum in order to measure the degree of severity of issues related to each authentication method.
  - **"Automation versus Human"** range from 1 = Human is better to 5 = Machine is better.
- **"Total Transaction Time (seconds)"** corresponds to the time it takes for a single user to present the biometric (acquisition time), processing time, and, optionally, might include entry of a PIN or user identifier (Woodward et al., 2003).
- **"Accuracy"** has two measure rates of authentication by biometrics:
  - False rejection rate (FRR) where a legitimate user is rejected by the acquisition device.
  - False acceptance rate (FAR) where a false user is accepted.
- **"Average Attack Space"** corresponds to the number of guesses made by an attacker in order to disclose the base secret (e.g. passwords, PINs, etc.).
- Abbreviations used are the following:
  - C/R = Challenge/Response
  - PK = Public Key
  - PRK = Private Key
  - SSO = Single-Sign-On
  - TGS = Ticket Granting Service.

Table 3.3 Comparative Analysis of User Authentication Methods

Attributes	Passwords (PW)	PIN	Proximity Card	One Time Password	C/R	Chip Card	Public key	Kerberos	Fingerprint, Hand, Face	Voice	Signature	Retina or Iris	Keystroke	Underskin Chip
Classification	Knowledge-Based (4 to 8 digits).	Knowledge-Based (4 to 8 digits).	Authentication token. Acquisition device.	Authentication token; Acquisition device (7)	Authentication token. Random challenge (7).	Smart Card. Embedded integrated chip.	Public Key Cryptography PK and PRK (8)	Key Distribution Center (KDC) (15)	Biometrics	Biometrics	Biometrics	Biometrics	Biometrics: user's typing rhythm	Human chip-based
Advantages	Easily implemented. More robust if system generated.	Not sent across the network	Last longer, no physical contact (11).	Very difficult to guess: passcode (user PIN+code from server)	PW can be a PIN (4 digits), easier to remember.	More secure than the regular user ID and password	PK provides much stronger identity checking	Trusted third-party	Easily sampled & non-intrusive	People use instinctively	Highly accepted by users	Unchangeable during lifetime	Enrollment and verification don't bother the regular work flow	Forgery, stealing, or removing the chip is too difficult
Disadvantages	Can be forgotten. Users create easily identifiable base secrets	Can be forgotten	Forgery (Easy to copy), loss, theft possible	Brute-force and dictionary attacks	Users share their access permissions	Need a smart card reader	Key distribution/ exchange	Scalability (centralized administration)	Criminal affiliation	Changes over time	Signature can change at any time	Requires much user cooperation	Impersonation attack	Impersonation attack
Characteristics/ Authentication Methods (or Data Generator)	Passwords (PW)	PIN	Proximity Card	One Time Password	C/R	Chip Card	Public Key	Kerberos	Fingerprint, Hand, Face	Voice	Signature	Retina or Iris	Keystroke	Underskin Chip
Security (Minimum=1 Maximum=5)	2	2	3	4	3	4	4	4	4	1	3	4	3	4
Costs (CADS)	$110 Password Cost Calculator (1)	$110 Password Cost Calculator (1)	$30 Card	$80 Token	$80 Token	$80 Card	$220 per seat for around 5000 users	Free software	$100 to $500 acquisition device	$5 acquisition device	$300 (Smith, 2002) acquisition device	$300–700 (Smith, 2002) acquisition device	$45 (keyboard) Available free software	$300 Chip
Usability (Minimum=1 Maximum=5)	1 Break 6 rules of UI design (Schneiderman, 1998).	2 More robust if system generated.	4 Practical for users (2)	2	2	3 Used in combination with PIN (2)	2 (2) (PGP)	2 SSO	4 (10)	4 Neural Net (3) (10)	4 (10)	4 (10)	4 (10)	4 (10)
Human Versus Automation (15) (Human is better=1; Machine is better=5)	5 Computer generates more secure, automatic passwords than humans.	5 Computer generates automatic passwords than humans.	5	5	5	5	5	5	1	1	1	1	1	1

(Continued)

Table 3.3 (Continued) Comparative Analysis of User Authentication Methods

Attributes	Passwords (PW)	PIN	Proximity Card	One Time Password	C/R	Chip Card	Public key	Kerberos	Fingerprint, Hand, Face	Voice	Signature	Retina or Iris	Keystroke	Underskin Chip
Characteristics/ Authentication Methods (or Data Generator)	**Pass Words (PW)**	**PIN**	**Proximity Card**	**One Time Password**	**C/R**	**Chip Card**	**Public Key**	**Kerberos**	**Fingerprint, Hand, Face**	**Voice**	**Signature**	**Retina or Iris**	**Keystroke**	**Underskin Chip**
Total Transaction Time (in seconds) (Woodward et al., 2003)	5–15	5–8	2–6	10–20	10–15	5–15	5–15	5–15	Face = 10–15; Fingerprint = 2–9; Hand = 4–10	10–12	10–15	4–12	10–15	3–6
Industrial Application	Unix (Smith, 2002), Windows NT/2000, Mac Keychain	RSA SecurID authenticators (4)	XyLoc (12)	RSA SecurID® 800 Authenticator	Safe Word, Crypto Card, Active Card	Gemalto (16)	Pretty Good Privacy (Symantec) (5)	Kerberos v.5 1.13.2 (15)	Apple TouchID iPhone 6 (fingerprint) NeoFace® Smart ID NEC (face)	Apple Mac OSX, Voice Security	Cyber-SIGN	IrisID	KeyLog PC (6)	No commercial implementation yet.
Accuracy	Average Space Attack: Dictionary attack = 2^{18} to 2^{20} (Smith, 2002)	13-bit (Smith, 2002)	Fair (no available quantitative data)	Average Space Attack = 2^{39} to 2^{63} (Smith, 2002)	Average Space Attack = 54 bits	Distance and cycle delay do have an impact on the accuracy of Read/Write.	Average Space Attack = 1024-Bit Public Key = 2^{86} (Smith, 2002) (13)	Requires clock synchronization between machines on the network (9)	FR = 3 to 7 in 1,000 (0.3–0.7%); FA = 1 to 10 in 100,000 (0.001–0.01%) (Bolle et al., 2004)	FR = 10 to 20 in 100 (10–20%); FA = 100 to 1000 in 100,0000 (Bolle et al., 2004)	FR = 2-try: 2.10%; FA = 2-try: 0.58%. (Bolle et al., 2004)	FR = 2 to 10 in 100 (2–10%); FA = 10 to 5 (0.001%) (Bolle et al., 2004)	No available data	No available data

Notes are displayed as numbers between parentheses (e.g. (1), (2), and so on), and can be found right after the comparative analysis table. They are not displayed in order of entry, given that different items may use the same note. References are also used due to the variety of data provided and the need to minimize the number of notes and/or footnotes on the pages. They can be found in the "References" section.

[1] The Password Cost Estimator shows the direct and recurring costs to your organization regarding the use of passwords. Although passwords seem to be free, they actually cost organizations a significant portion of its IT support budget. This silent budget killer is merely the time the technical support personnel devotes to resetting users' passwords even though self-service reset password systems are being employed more and more. This does not include the abstract costs associated with lost productivity of users, security breaches, etc. but only the labor cost of the help desk personnel physically resetting passwords on the system. Enabling Compliance with Password Policies, Password Cost Calculator. Mandylion Research Labs, LLC. March 30, 2016 <http://www.mandylionlabs.com/PRCCalc/PRCCalc.htm>

[2] Average swiping speed. The ideal swiping speed deals with your self-confidence: shy people swipe slower, anxious people swipe fast, and confident people swipe at the ideal speed.

[3] Novel Neural Net Recognizes Spoken Words Better than Human Listeners. October, 2015 <http://www.sciencedaily.com/releases/1999/10/991001064257.htm>

[4] RSA SecurID authenticators. RSA Security. August, 2018 <https://community.rsa.com/docs/DOC-40601#token>

[5] Pretty Good Privacy (PGP) Symantec Desktop Email Encryption program from Symantec. October, 2015 <http://www.symantec.com/encryption/>

[6] Verification is built up on the concept that the rhythm with which the user types is distinguishing. KeyLog PC Key Stroke Logger. August, 2018 <https://www.logix-oft.com/en-ca/index>

[7] RSA hardware authenticators. RSA Security. August, 2018 <https://community.rsa.com/docs/DOC-40601#token>

[8] To reduce risk exposure and comply with security regulations, companies rely on Public Key Infrastructure (PKI) and digital certificates. 10/05/2015 <https://en.wikipedia.org/wiki/Public-key_cryptography>

[9] Maximum tolerance for computer clock synchronization: This is the maximum time that can be tolerated between a ticket's timestamp and the current time at the Kerberos Distribution Center (KDC). Kerberos configuration by Jan De Clercq, October 8, 2004, Elsevier Digital Press

[10] User data collection can impact usability as well. User data collection is the time period a person must spend to have his/her biometric reference template successfully created (i.e. enrollment and verification time) but can vary dramatically depending on the biometric device.

[11] Cards are intended to operate within up to 10 cm of the reader antenna at a frequency of 13.56 MHz. ISO/IEC 14443-1:2008 Identification cards – Contactless integrated circuit cards – Proximity cards. Part 1: Physical characteristics. <http://www.iso.org/iso/home/store/catalogue_ics/catalogue_detail_ics.htm?csnumber=39693>.

[12] XyLoc provides full-time access control by determining a user's location and automatically locking the computer when the user is not physically present. Ensure Technologies Inc. <http://ensuretech.com/wp-content/uploads/2011/08/xyloc-overview-brochure.pdf>

[13] Cost of cryptographic operations (1,280-bit Rabin-Williams keys on 550 MHz K6). Mazieres, D. & Zhu, Y.: 2002. Machine Learning course - G22.3033-001. Topics in Computer System Security, New York University, NY (USA):

Operation	Time (seconds)
Encrypt	1.11
Decrypt	39.62
Sign	40.56
Verify	0.10
Total	81.39

[14] Form of identifier presented by the user: PIN, memory card, etc.

[15] MIT Kerberos Distribution page, v 1.286, MIT Kerberos <http://web.mit.edu/kerberos/dist/>. Kerberos V5 Release 1.13.2 - current release, 2015-02-11 http://web.mit.edu/kerberos/dist/#krb5-1.13

[16] Smart card readers, 05/10/2015 <http://www.gemalto.com/readers>

Chapter 4

Fundamentals of the Usable Security Protocol for User Authentication

Summary

This chapter details the most representative usability inspection methods currently used in the human–computer interaction (HCI) landscape. The chapter paves the ground for the proposed protocol, which will be detailed in the next chapters. These methods serve as a foundation (and data source) for the development of the Usable Security Symmetry inspection method, which will be described in the next chapters. It is worth noting that due to the absolute lack of specific standards and guidelines for *user authentication*, human–computer interaction (security) (HCISec) research, in turn, will be described, including both previous and current work related to usable security of computer security mechanisms. As a matter of fact, it was very beneficial in researching usable security related to a broader spectrum of security mechanisms; firstly, because of the referred to lack of user authentication guidelines, and secondly, to understand what these security mechanisms are, how they work, and verify if any of these guidelines could be used and adapted for user authentication.

4.1 Introduction

Engineering models for human performance allow some aspects of user interface design to be evaluated analytically for usability (without consuming resources for empirical user testing) by making usability predictions based on an analysis of the user's task in conjunction with principles and parameters of human performance (John and Kieras, 1994). Broadly speaking, engineering models predict the time required to complete a task, the demands on the user's memory, and the respective learning required to make use of the design.

The overall scheme for using engineering models in the user interface design process is briefly described in the following steps:

1. Develop an initial task analysis.
2. Propose the first interface design.
3. Use an engineering model as applicable to find the usability problems in the interface.
4. Perform a user testing. (This is done because there are other aspects of usability that are poorly understood, and some form of user testing is still required to ensure a quality result. Only after dealing with design problems revealed by the engineering model would the designer then go on to user testing step.)
5. Review the design if the user testing reveals a serious problem(s). The engineering model will help refine the redesign quickly. Thus, the slow and expensive process of user testing is reserved for those aspects of usability that can only be addressed at this time by empirical trials.

If engineering models can be fully developed and put into use, the designer's creativity and development resources can be more fully devoted to more challenging design problems, such as creating entirely new interface concepts or approaches to the design problem at hand.

4.2 The Goals, Operators, Methods, and Selection Rules (GOMS) Model

The major existing form of an engineering model for interface design is the Goals, Operators, Methods, and Selections rules (GOMS) model, first proposed by (Card et al., 1983). GOMS is a family of predictive models of human performance that can be employed to refine the efficiency of HCI. It is a widely employed method by usability specialists for computer system designers because it produces quantitative and qualitative predictions of how people will use a proposed system.

GOMS is a technique for task analysis, and it is based on iterative testing and design revision using actual users to test the system and help identify usability problems. It is broadly agreed that this approach, derived from human factors, does indeed work when carefully applied (Card, 1983). The GOMS model is appropriate for modern software development practice due to being faster and cheaper than empirical user testing, especially when domain experts and target users are scarce.

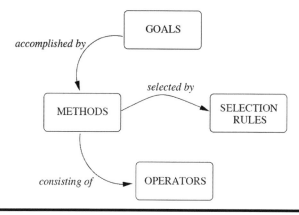

Figure 4.1 The Goals, Operators, Methods, and Selections rules (GOMS) model (Card et al., 1983).

The GOMS model is the general term for a family of four human information processing techniques such as keystroke-level model (KLM), Card, Moran, and Newell GOMS (CMN-GOMS), Natural GOMS Language (NGOMSL), and Cognitive-Perceptual-Motor GOMS (CPM-GOMS). These techniques range from the trivial, in the case of the KLM, to slightly more complex for CMN-GOMS, to an elaborated sequential architecture with a working memory and specified procedure knowledge representation in NGOMSL, and finally to a powerful but relatively unspecified multiple parallel processor architecture in CPM-GOMS. The NGOMSL technique will be described in the next sections, whereas the other ones are beyond scope of this book and can be found in John & Kieras (1996a).

As mentioned, the GOMS model attempts to model and predict user behavior as illustrated in Figure 4.1. It represents users' knowledge in a hierarchical stack structure consisting of *Goals*, the state of affairs to be achieved; *Operators*, elementary perceptual, motor, cognitive acts whose execution is necessary to change any aspect of a user's mental state or to affect the task environment; *Methods*, a series of steps consisting of operators that the user performs to accomplish a goal. A method may call for sub-goals, so the methods have a hierarchical structure. If there is more than one method to accomplish a goal, then *Selection Rules* choose the appropriate method, depending on the context. Describing the goals, operators, methods, and selection rules for a set of tasks in a formal way constitutes doing a GOMS analysis or constructing a GOMS model.

4.2.1 GOMS: A Method for Cognitive Task Analysis

Advances in technology have been significantly increasing the demands on the cognitive skills of workers, whereas the human system has remained moderately stable. Workers have been employing and/or operating complex computer technologies.

User interfaces (UIs) help users interact with programs and in their tasks – users employ programs to perform their tasks. A UI should not reflect the structure of the underlying program, but the structure of the task domain. Therefore users should not interact with the computer, but with their tasks.

Cognitive task analysis (CTA) can boost human performance by guiding the development of tools and programs that support the cognitive processes required for a task (Chipman et al., 2000). CTA is conducted for a wide variety of purposes, such as system development, instruction and training, and human–computer interface design. For the particular research of this book, CTA provides a description of the conceptual and procedural knowledge used by users as they perform, for example, authentication tasks (e.g. accessing a protected network resource using a knowledge-based authentication (KBA) method).

According to Proctor & Vu, (2005) (Williams & Voigt, 2004), GOMS is the most widely used and most accurate method of cognitive task analysis. GOMS is both a performance model and a cognitive task analysis method. Some of the types of applications in which cognitive task models have been applied in their research include assessing human–computer interaction complexity, determining the productivity of human–computer interfaces, and analyzing an interface design to determine whether methods can be automated.

A GOMS task analysis is a process of creating a GOMS model by decomposing user task knowledge into GOMS components. The general strategy is similar to hierarchical task analysis: Begin by identifying the top-level user goals, emphasizing breadth over depth. Then refine each goal into sub-goals, methods, and operators. It differs from hierarchical task analysis in that a GOMS model has a specific format. GOMS is specifically designed to represent procedural knowledge of well-learned cognitive tasks. Additionally, the GOMS model can be translated into executable programs for evaluating the consistency of the model and obtaining quantitative measures of the interface being designed.

In the analysis and implementation stages, a GOMS model of existing or new tasks can be used to determine if the functional requirements are derived from the tasks performed by users; every task goal should have a specific method for achieving the goal. The GOMS analysis can identify benchmark tasks (i.e. a set of benchmark cases that represent important user tasks) and establish performance criteria for user testing at later stages. A GOMS model can also be used to determine the consistency of procedures: similar goals should be accomplished with similar tasks. When more than one method is provided, the GOMS analysis can determine whether there are selection rules that determine the method to be used. When a GOMS model is built, quantitative assessments of the design can be made from assumptions about operator execution times. For example, a GOMS model can be used to evaluate alternative design concepts in terms of time to learn to use the system, time to perform some tasks, and possibly time for recovering from errors.

GOMS analyses and models are most appropriate for user tasks that have well-defined goals (e.g. log on to your online banking website) and that require the application of learned cognitive skills. GOMS also does not permit analysis of interface issues related to the layout of components, readability of text, and so on. Instead, it is focused on the procedures that the user must learn and execute when performing tasks; GOMS represents only the procedural aspects of usability. GOMS models can predict the procedural characteristics of usability; these characteristics concern the amount, consistency, and effectiveness of the procedures that users must pursue. Since the usability of numerous systems depends profoundly on the simplicity and effectiveness of the procedures, the GOMS model has significant value in guiding interface design. The reason why GOMS models can predict these characteristics of usability is that the methods for achieving user goals have a tendency to be strongly constrained by the design of the interface, making it possible to build a GOMS model given just the interface design, prior to any prototyping or user testing (Kieras, 1996).

Clearly, there are other important characteristics of usability that are not related to the procedures entailed by the interface design. These characteristics concern both lowest-level perceptual issues like the font's readability on cathode ray tubes (CRTs)*, and also very high-level issues such as the user's conceptual knowledge of the system (e.g. whether the user has an appropriate mental model) (Kieras and Bovair, 1984), or the degree to which the system fits properly into an organization (i.e. the social or organizational impact of the system and the resulting influence on productivity) (John and Kieras, 1994). The lowest-level issues handle well in terms of standard human factors methodology, while understanding the higher-level concerns is at present an issue of good judgment on the part of the practitioner and the higher-level task analysis techniques (Kieras et al., 1995).

Based on extensive research among a variety of current CTA strategies (Kirwan et al., 1992; Hollnagel, 2003; Kieras et al., 1995; Diaper and Stanton, 2003; Hackos and Redish, 1998), and in the context of the topic under consideration, user authentication, the NGOMSL (Natural Goals, Methods, Selection Language) (Kieras, 1996) was selected as the most appropriate CTA method.

4.2.2 How to Develop a GOMS Model

In all GOMS analysis techniques, the analyst must start with a list of high-level user goals. Typically, this list of goals can be obtained from other task analyses, including observations of users of similar or existing systems. The GOMS steps are described in detail in the next paragraphs.

4.2.2.1 Identify User's Goals

The analyst can express in a GOMS model how the user can accomplish these goals with the system being designed. A goal is something that the user tries to

---

* CRT is the technology used in most televisions and computer display screens.

accomplish. The analyst attempts to identify and represent the goals that typical users will have. A set of goals will usually have a hierarchical structure in which accomplishing a goal may first require accomplishing one or more sub-goals. A goal description is an action–object pair in the form <verb noun> (e.g. access a file). The verb can be complicated if it is necessary to distinguish between methods (e.g. "move-by-find-function cursor"). Any parameters or modifiers, such as where a "to-be-deleted" word is located, are represented in the task description.

4.2.2.2 Define Methods

Write a method to accomplish a goal – it may invoke sub-goals. The example consists of a list of methods for each system. For example: Select a word expressed in the NGOMSL notation is shown in Figure 4.2.

4.2.2.3 Define Operators

Standard primitive external operators, standard primitive mental operators, and *analyst-defined mental operators* are mostly determined by the hardware and lowest-level software of the system (e.g. move finger to the USB fingerprint reader). *External operators* are the observable actions through which the user exchanges information with the system or other objects in the environment (e.g. a perceptual operator such as "Read your online ID" in the screen). *Mental operators* are the internal actions performed by the user. In the notation system presented here, some mental operators are "built in". These operators correspond to the basic mechanisms of the cognitive processor, the cognitive architecture.

A particular task analysis assumes a particular level of analysis which is reflected in the "grain size" of the operators. If an operator will not be decomposed into a finer level, then it is a primitive operator. But if an operator will be decomposed into a sequence of lower-level, or primitive, operators, then it is a high-level operator. Exactly which operators are primitives depends on the finest grain level of analysis desired by the analyst. Some typical primitive operators are actions like pressing a button or moving the hand. All built-in mental operators are primitive. High-level operators would be gross actions, or stand-ins for more detailed analysis, such as LOG-INTO-SYSTEM. The analyst recognizes that these could be decomposed, but may choose not to do so, depending on the purpose of the analysis (Kieras, 1996).

Method for goal: Select word
Step 1. Locate middle of word.
Step 2. Move cursor to middle of word.
Step 3. Double-click mouse button.
Step 4. Verify correct text is selected.
Step 5. Return with goal accomplished.

Figure 4.2   Method for selecting a word (Kieras, 1996).

In the case of the *standard primitive external operators*, the designer defines the primitive motor and perceptual operators based on the basic actions required by the system being analyzed. These correspond straightforwardly to the physical and some of the mental operators used in the keystroke-level model. The KLM uses only keystroke-level operators, no goals, methods, or selection rules. The analysis simply lists the keystrokes, mouse movements, and mouse-button presses that a user must perform to accomplish a task, and then uses a few simple heuristics to place a single type of trivial "mental operator" which approximates many kinds of internal cognitive actions (e.g. thinking).

Standard primitive mental operators are in turn divided into *flow of control* and *memory storage and retrieval*. *Flow of control* represents a sub-method which is invoked by declaring its goal: ACCOMPLISH GOAL: <GOAL DESCRIPTION>. Control passes to the method for the goal and returns here when the goal has been accomplished. The operator RETURN WITH GOAL ACCOMPLISHED marks the end of a method. A decision is represented by a Decide operator which contains either one IF-THEN conditional with an optional ELSE, or any number of IF-THEN conditionals. In *memory storage and retrieval,* the memory operators reflect the distinction between long-term memory (LTM) and working memory (WM). They are typically used in computer operation tasks as follows:

Recall that <WM-object-description>
Retain that <WM-object-description>
Forget that <WM-object-description>
Retrieve-from-LTM that <LTM-object-description>

Finally, *analyst-defined mental operators* represent psychological processes that are too complex to indicate as methods in the GOMS model, so the designer can circumvent these processes by defining operators that act as placeholders for the mental activities. They are mostly high-level operators. For example, THINK-OF <DESCRIPTION> represents a process of thinking of a value for some parameter designated by <DESCRIPTION> and putting the information into working memory (Kieras, 1996).

4.2.2.4 Selection Rules

The purpose of a selection rule is to route control to the appropriate method to accomplish a goal (Kieras, 1996). If there is more than one method for a goal, then a selection rule is logically required as shown below:

Selection rule set for goal: <general goal description>
If <condition> Then accomplish goal: <specific goal description>.
If <condition> Then accomplish goal: <specific goal description>.
. . .
Return with goal accomplished.

A practical example for "Close a Window" scenario using GOMS model is shown below:

GOAL: CLOSE-WINDOW
      . (select GOAL: USE-MENU-METHOD
          . MOVE-MOUSE-TO-FILE-MENU
          . PULL-DOWN-FILE-MENU
          . CLICK-OVER-CLOSE-OPTION
          . GOAL: USE-CTRL-W-METHOD
              . PRESS-CONTROL-W-KEYS)

For a particular user:
Rule 1: Select USE-MENU-METHOD unless another rule applies
Rule 2: If the application is GAME,
                select CTRL-W-METHOD

4.2.3 Natural GOMS Language (NGOMSL)

In psychology, researchers fit the parameters of their models to data they have collected on the task they are studying. But in interface design, system developers need quantitative a priori predictions for systems that have not yet been built. Thus, HCI researchers have done extensive theoretical and empirical work to estimate parameters that are robust and reliable across tasks and can be used *without further empirical validation* to make predictions (e.g. usability testing). NGOMSL does quantitative a priori predictions for systems that have not yet been built.

NGOMSL is a structured natural language notation for representing GOMS models and a procedure for constructing the same models (Kieras, 1996). An NGOMSL model is in program form and provides predictions of operator sequence, execution time, and time to learn the methods. An analyst constructs an NGOMSL model by performing a top-down, breadth-first expansion of the user's top-level goals into methods until the methods contain only primitive operators, typically keystroke-level operators (e.g. click on Sign In button with left mouse button). Like CMN-GOMS, NGOMSL models explicitly represent the goal structure and can therefore represent high-level goals.

The NGOMSL technique refines the basic GOMS concept by representing methods in terms of a cognitive architecture called cognitive complexity theory (CCT) (Kieras and Polson, 1985). This cognitive theory allows NGOMSL to incorporate internal operators such as manipulating working memory information or setting up sub-goals. Because of this, NGOMSL can also be used to estimate the time required to learn how to achieve tasks. An example of an NGOMSL model taken from John and Kieras (1996a) is shown in the following.

Method for goal: Cut text

Step 1. Accomplish goal: Highlight text.

Step 2. Return that the command is CUT, and accomplish goal: Issue a command.

Step 3. Return with goal accomplished.

...

Selection rule set for goal: Highlight text

If text is word, then accomplish goal: Highlight word.

If text is arbitrary, then accomplish goal: Highlight arbitrary text.

Return with goal accomplished.

...

Method for goal: Highlight arbitrary text

Step 1. Determine position of beginning of text (1.20 sec)

Step 2. Move cursor to beginning of text (1.10 sec)

Step 3. Click mouse button. (0.20 sec)

Step 4. Move cursor to end of text. (1.10 sec)

Step 5. Shift-click mouse button. (0.48 sec)

Step 6. Verify that correct text is highlighted (1.20 sec)

Step 7. Return with goal accomplished.

This NGOMSL model predicts that it will take 5.28 seconds to highlight arbitrary text.

4.2.3.1 Cognitive Complexity Theory

When you have learned to perform a task with a particular interface and have to switch to doing the same task with a new interface, how much better off will you be than someone just learning to do the task with the new interface? That is, how much is the knowledge gained from using the old interface "transferred" to using the new interface? Cognitive complexity theory (CCT) (Bovair et al., 1990; Kieras and Polson, 1985) is a psychological theory of transfer of training applied to HCI. It seeks to decompose user goals for completing computer tasks with a greater degree of granularity than GOMS in order to obtain more accurate predictions of how long it will take users to learn to complete tasks online with fewer errors. In contrast to GOMS models, CCT investigates learners rather than skilled users. CCT has been shown to provide good predictions of execution time, learning time, and transfer of procedure learning.

CCT assumes a simple serial stage architecture in which working memory (WM) triggers production rules that apply at a fixed rate. These rules alter the contents of working memory or execute primitive external operators such as making a keystroke. GOMS methods are represented by sets of production rules in a prescribed format. Learning procedural knowledge consists of learning the individual production rules. Learning transfers from a different task if the rules have already been learned. The complexity of a task will be reflected in the number and content of the production rules. The time it takes to learn a task is a function of the number of new rules that the user must learn; if the user already has a production rule, and a new task requires a rule that is similar, then the rule for the new task need not be learned. It is important to note that some predictions about errors and speedup, with practice, can also be collected from the contents of the production rules.

The association between the NGOMSL notation and the CCT architecture is in fact direct: There is basically a 1:1 relationship between statements in the NGOMSL language and the production rules for a GOMS model written in the CCT format. Therefore, the CCT prediction results can be used by NGOMSL models to estimate not only execution time like KLM and CMN-GOMS* but also the time to learn the procedures.

The CCT and NGOMSL models have been empirically validated at the KLM of analysis (operators like DETERMINE-POSITION and CLICK-MOUSE-BUTTON). Therefore, models at that level can generate trustworthy quantitative estimates.

Because NGOMSL models specify methods in program form, they can characterize the procedural complexity of tasks – both in terms of how much must be learned and how much has to be executed. CCT employs production systems in the form IF (condition) THEN (action) in order to explain the cognitive demands related to task performance. The condition component of production rules relates to either the contents of WM or environmental factors (e.g. screen display). The action component relates to manipulations of either the environment (e.g. key presses) or the contents of WM (e.g. deleting current goals). The clauses included within the "condition" component, are combined using logical AND. If the pattern of goals, notes, and external information in WM matches the condition clauses, the rule is said to "fire" and the action operators are executed. Current goals and variables have to be stored in WM, and it is on this basis that task demands are estimated. Once added to WM by a production, GOALS and NOTES must be preserved in WM until deleted by later productions. A number of production systems may be produced with the purpose of describing complete task performance. The sequence in which these productions are performed may depend upon selection rules that specify different methods of achieving the current task goals (Card et al., 1983).

To illustrate a CCT production rule, consider the following example:

Editing with VI (Visual editor)[†]:

---

* CMN-GOMS adds hierarchical structure to the KLM version of GOMS. Tasks are organized as a series of goals and sub-goals and operators are organized into subroutines called methods. CMN-GOMS can provide task execution times and afford a better view of the task structure than KLM.

[†] VI (visual editor) is available on all major computer systems. VI is a display oriented, interactive text editor which allows a user to create, modify, and store files on the computer via a terminal.

- Production rules are in LTM.
- Model working memory as attribute-value mapping:
  (GOAL perform unit task)
  (TEXT task is insert space)
  (TEXT task is at 5 23)
  (CURSOR 8 7)
- Rules are pattern-matched to WM.
  - e.g. LOOK-TEXT task is at %LINE %COLUMN is true, with LINE = 5
    COLUMN = 23.

4.2.3.2 NGOMSL Steps Development Process

This section lists the steps involved in the development of the NGOMSL model. Each step contains information on which processors its operators require. *Interstep mental operators* (e.g. Recall_Passcode) require the use of the cognitive processor, but *intrastep mental operators* (e.g. Return_with_goal_accomplished) do not. The following describes the required steps:

- Generate task description;
- Describe a list of high-level user goals;
- Define operators, write methods, and selection rules for accomplishing goals;
- Estimate pure-learning time.

4.2.4 Learning Time Predictions

NGOMSL models have been demonstrated to be superior predictors of the time it takes to learn how to use a system, keeping in mind that what is predicted is the pure learning time (PLT) for the *procedural knowledge* represented in the methods. As already mentioned, the user is assumed to already know how to execute the operators; the GOMS methods do not represent the knowledge involved in executing the operators themselves but rather only represent the knowledge of which operators to apply and in what order to accomplish the goal. Innovative interface technology frequently results in new operators; moving the cursor with a mouse was a new operator, and selecting objects with an eye-movement tracker or manipulating 3D objects and flying about in virtual space with data-glove gestures will be new operators, as these technologies are already moving into the workplace. Undoubtedly, the time to learn how to execute new operators is a decisive aspect of the value of new interface devices, but a GOMS model that *assumes* such operators cannot predict their learning time. The time to learn new operators themselves would have to be measured or simply not be included in the analysis.

The total elapsed time to learn to use a system depends not simply on how much procedural knowledge must be learned but on how much time it takes to complete the training curriculum itself. It means that nearly all learning of computer use occurs in the context of the new user performing tasks of some sort, and this performance would take a certain amount of time even if the user were fully trained. As a result, the total learning time includes the time to execute the training tasks in addition to the extra time required to learn how to perform the tasks, the PLT. As Gong (1993) has demonstrated, training task execution times can be predicted from a GOMS model of the training tasks.

The fundamental empirical result is that the time needed to learn a particular procedure is roughly linear with the number of NGOMSL statements that must be learned. Thus, the PLT for the methods themselves can be estimated just by counting the statements and multiplying by an empirically determined coefficient. The transfer of training effects can be calculated by deducting the number of NGOMSL statements in methods that are identical, or highly similar, to ones already known to the learner (Bovair et al., 1990). This description of interface consistency in terms of the quantitative transferability of procedural knowledge is possibly the most important contribution of the existing CCT research and the NGOMSL technique.

A supplementary element of the PLT is the time required to memorize chunks of declarative information required by the methods, such as the menu names under which commands are found. Such items are assumed to be stored in LTM and, while not rigorously part of the GOMS methods, are required to be in LTM for the methods to execute correctly. Including this component in learning time estimates is an approach to representing the learning load imposed by menu or command terms.

"The validity and utility of the learning time estimates depend on the general requirements of the learning condition. Obviously, if the learner is engaged in problem-solving, or in an unstructured learning situation, the time required for learning is more variable and ill-defined than if the learner is trained in a tightly controlled situation" (John and Kieras, 1996a). Also, it seems logical that in spite of the learning situation, systems whose methods are longer and more complex will require more time to learn them because more procedural knowledge has to be acquired, either by explicit study or inferential problem-solving. Summarizing this discussion of estimating learning using the values determined, Gong (1993) describes the following:

Total Learning Time = Pure Procedure Learning Time *
+Training Procedure Execution Time

*Pure Procedure Learning Time = NGOMSL Method Learning Time + LTM Item Learning Time.

These formulas give a pure procedure learning time estimate for a whole set of methods in a usual learning situation, assuming no previous knowledge of any methods and assuming that learning the appropriate command words for the two menu terms will require learning three chunks each (John and Kieras, 1994).

4.2.5 Execution Time Predictions

As already mentioned, the execution time for a task is predicted by simulating the execution of the methods required to perform the task as follows:

> **Execution Time = NGOMSL statement time**
> **+Primitive External Operator Time**
> **+Waiting Time**

The execution time predictions are founded on the sequence of operators executed while carrying out the benchmark tasks. Execution time might be approximated by a constant, by a probability distribution, or by a function of some parameter. For example, the time to type a word might be approximated by a constant (e.g. the average time for an average typist to type an average word), or a statistical distribution, or by a function involving the number of letters in the word and the time to type a single character (which could, in turn, be approximated by a constant or a distribution) (John and Kieras, 1994). The precision of estimates obtained from a GOMS model depends on the precision of this assumption and on the precision of the duration estimates.

4.2.6 NGOMSL Methodology

According to Kieras (2006), the NGOMSL methodology comprises of the following steps:

- Top-down breadth-first task decomposition:
  - Start with the user's top-level goals.
  - Write a step-by-step procedure for accomplishing each goal in terms of sub-goals or keystroke-level operators.
  - Use NGOMSL syntax for the procedure.
  - Recursively write a method for each sub-goal until all methods contain only keystroke-level operators.
  - Write a selection rule to specify which method to use if more than one for a goal.
- Count the number of statements in methods to predict learning time:
  - Consistency can be directly reflected by the presence of reused sub-methods, reducing learning time.
  - Similar methods also reduce learning time.

■ For a specific task scenario, count the number of statements and operators executed to predict execution time.

4.2.7 GOMS Limitations

The GOMS model has a number of limitations, such as that the predictions are only valid for expert users who do not make any errors. This is, in fact, a significant deficiency because even expert users do make mistakes. Given that one of the goals of HCI is to seek maximum usability for all users – including novices – this is an insufficiency in the GOMS model. However, the only GOMS technique that mitigates this deficiency is exactly NGOMSL, which attempts to model the time required to learn a task. As already mentioned, the GOMS model has been used by the authors as a cognitive task analysis tool and not for evaluating a specific user interface. Therefore, the goal here is not to measure the user performance related to user authentication methods but rather in identifying and understanding the cognitive processes involved specifically in those types of interactions.

Another limitation is that NGOMSL shapes all tasks as goal-directed, neglecting the problem-solving nature of some tasks. It does not take into account individual differences among users. One of the main premises of the authors is that tasks should be goal-directed (e.g. make an online funds transfer) and the user authentication component should be present in them (e.g. sign in to online banking). This goal directness "capability" is in fact very suitable to the type of tasks described in this book. Computer security applications, especially user authentication applications, usually have a clear-cut, narrow user interface and function-based overall structure. This entails limited application features to users, but, at the same time, simplicity is gained.

GOMS models cannot provide information on how valuable or pleasant the product under design will be. Finally, as already mentioned, GOMS does not address the social or organizational impact of the product under development.

4.3 Usability Evaluation Methods

Usability evaluation methods can be divided into three categories: *test, inspection,* and *inquiry*. Each category can be applied to one and/or more phases of the design lifecycle as follows:

■ *Test method.* This method makes use of representative users to work on typical tasks using a system (e.g. traditional usability testing), and their performance is usually measured.
■ *Inspection method.* In this approach, usability experts – and sometimes software engineers or domain experts – inspect usability-related aspects of a user interface (e.g. heuristic evaluation) (Nielsen, 1994a). Expert-based evaluations of usability are similar to design reviews of software projects and code

walkthroughs. Inspection methods include heuristic evaluation, guideline reviews, pluralistic walkthroughs, consistency inspections, standards inspections, cognitive walkthroughs, formal usability inspections, and feature inspections. One interesting characteristic of the inspection method when compared to the other categories is that they can be used at any stage of design, from product definition to final design. Usability inspection methods are particularly efficient regarding the high benefit to cost ratio and also able to find many usability problems that are overlooked by user testing (Karat et al., 1992), (Desurvire et al., 1992), (Desurvire, 1994).

∎ *Inquiry method*. This method collects information regarding users' preferences, desires, and behaviors (e.g. focus group), and aims to formulate the requirements of a design.

Currently, there is a multiplicity of methods used to evaluate usability. To choose a specific method, you must consider human and facilities resources, costs, time constraints, and the suitability of the method for the product at hand.

The following section presents an overview of the three most representative usability inspection methods relevant to the topic of user authentication: heuristic evaluation, cognitive walkthrough, and the GOMS model. These methods are also recognized as the most actively employed and researched usability inspection methods by the HCI research and industry communities, according to (Hollingsed and Novick, 2007) and (Kieras, 2006).

4.3.1 General Usability Principles ("Heuristics") for User Interface Design

One of the most widely recognized usability inspection methods, "heuristic evaluation"' is an inspection method employed to analyze the usability problems of a user interface or a system against a set of ten established principles called "heuristics" (Molich and Nielsen, 1990). These heuristics are more based on empirical data than on specific usability guidelines.

After evaluating numerous heuristics, Nielsen (1994a) developed the following ten heuristics:

∎ *Visibility of system status*: The system should always keep users informed about what is going on through appropriate feedback within a reasonable time.

∎ *Match between system and the real world*: The system should speak the user's language, with words, phrases, and concepts familiar to the user, rather than system-oriented terms. Follow real-world conventions, making information appear in a natural and logical order.

∎ *User control and freedom*: Users often choose system functions by mistake and will need a clearly marked "emergency exit" to leave the unwanted state without having to go through an extended dialogue. Support undo and redo.

- *Consistency and standards*: Users should not have to wonder whether different words, situations, or actions mean the same thing. Follow platform conventions.
- *Error prevention*: Even better than good error messages is a careful design which prevents a problem from occurring in the first place. Either eliminate error-prone conditions or check for them and present users with a confirmation option before they commit to the action.
- *Recognition rather than recall*: Minimize the user's memory load by making objects, actions, and options visible. The user should not have to remember information from one part of the dialogue to another. Instructions for use of the system should be visible or easily retrievable whenever appropriate.
- *Flexibility and efficiency of use*: Accelerators – unseen by the novice user – may often speed up the interaction for the expert user such that the system can cater to both inexperienced and experienced users. Allow users to tailor frequent actions.
- *Aesthetic and minimalist design*: Dialogues should not contain information which is irrelevant or rarely needed. Every extra unit of information in a dialogue competes with the relevant units of information and diminishes their relative visibility.
- *Help users recognize, diagnose, and recover from errors*: Error messages should be expressed in plain language (no codes), precisely indicate the problem, and constructively suggest a solution.
- *Help and documentation*: Even though it is better if the system can be used without documentation, it may be necessary to provide help and documentation. Any such information should be easy to search for, focused on the user's task, concrete in regards to the steps to be carried out, and not overly complex.

An example of such a (usability) heuristic:

Visibility of system status: Like how Twitter tells us when it's publishing a tweet and makes a chirping sound when it's done.

Heuristic evaluation is not limited to the aforementioned set of heuristics. Actually, it can be as long as the evaluators consider appropriate for the task at hand. For instance, you may create a specific list of heuristics for specific audiences, like senior citizens, children, or disabled users, based on a review of the literature (UPA, 2010).

Heuristic evaluation falls into the general category of usability inspection methods, and it is considered a "discount usability engineering" method (i.e. smaller and cheaper usability studies for projects with small budgets for usability).

Heuristic evaluation is performed individually by a group of three to five evaluators, and then results are presented to the team leader, who compiles a report of the findings. These evaluators should preferably be experts, who often possess the knowledge to design and carry out comprehensive performance tests with human subjects and the ability to analyze the resultant data. Their training is usually an advanced degree in human factors, behavioral science, industrial engineering, human–computer interaction, industrial design, computer science, or a related field. The greatest results are achieved by evaluators who are usability specialists with domain knowledge (e.g. computer security). Independent research has found heuristic evaluation to be extremely cost-efficient, confirming its value in circumstances where limited time or budgetary resources are available. "Overall, the heuristic evaluation technique as applied here produced the best results. It found the most problems, including more of the most serious ones, than did any other technique, and at the lowest cost" (Jeffries et al., 1991).

Another important factor to be taken into consideration is the ideal number of evaluators to perform a heuristic evaluation. According to Nielsen (1994a), it is normally recommended to use three to five as an optimum number of evaluators to ensure that the majority of usability problems will be found. The reason why is that there will be no such important gain by using larger numbers as shown in Figure 4.3.

The plotted curve shows how many times the benefits are greater than the costs for heuristic evaluation of a sample project using the assumptions discussed in Nielsen (1994b). The optimal number of evaluators in this example is four, with benefits that are 62 times greater than the costs. This curve provides the general observation that heuristic evaluation seems to work best with three to five evaluators. In the example illustrated in Figure 4.3, a heuristic evaluation with four evaluators would cost $6,400 and would find usability problems worth $395,000.

With regards to expert review versus usability testing, Molich and Dumas (2008) argue that "expert reviews with highly experienced practitioners may be

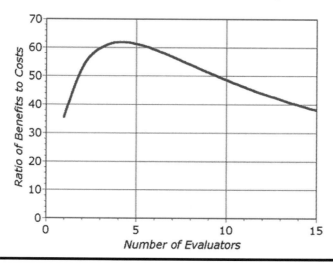

Figure 4.3 **The optimal number of usability evaluators: benefits versus costs (Nielsen, 1994b).**

more efficient than usability tests, in terms of number of issues found as a function of resources expended."

The next paragraphs describe the required procedures to perform a heuristic evaluation according to UPA (2010):

- Decide which aspects of a product and what tasks you want to review. For most products, you cannot review the entire user interface, so you need to consider what type of coverage will provide the most value.
- Decide which heuristic(s) will be used among the available ten heuristics.
- Select a team of three to five evaluators (you can have more, but the time to aggregate and interpret the results will increase substantially) and give them some basic training on the principles and process.
- Create a list of representative tasks for the application or component you are evaluating. You might also describe the primary and secondary users of your product if the team is not familiar with the users.
- Ask each evaluator to perform the representative tasks individually and list where the product violates one or more heuristics. After the evaluators work through the tasks, they are asked to review any other user interface objects that were not directly involved in the tasks and take notes of violations of heuristics. You may also ask evaluators to rate how serious the violations would be from the users' perspective.
- Compile the individual evaluations and ratings of the seriousness of the violations.
- Categorize and report the findings so they can be presented effectively to the product team.

The advantages of heuristic evaluation are that it is inexpensive, intuitive (i.e. we apply a set of predefined rules/heuristics), very easy to plan, and can be used (and feedback can be obtained) early in the design process. Assigning the right heuristic can aid in recommending the best corrective measures to the designer. The disadvantage is that there is a focus on problems rather than on solutions. Also, if the wrong heuristics are assigned to potential problems, it will mislead designers into applying the wrong solutions to the problems.

4.3.2 Cognitive Walkthrough

Cognitive walkthroughs are performed at any stage of design using a prototype, a conceptual design document, or the final product. This is a more specific version of a design walkthrough, focusing on cognitive principles. Based on a user's goals, a group of evaluators steps through tasks, evaluating at each step how difficult it is for the user to identify and operate the interface element most relevant to their current sub-goal and how clearly the system provides feedback for that action.

Cognitive walkthroughs take into consideration the user's thought processes that contribute to decision making, such as memory load and the ability to reason. For example, searching for the User Experience Professionals' Association website can be broken down to several levels of tasks. At a general level, it requires opening up a browser, remembering the Uniform Resource Locator (URL), and typing it in the browser's address bar. Or, if you do not remember the URL, you must choose a search engine, think of a search keyword, type the keyword in the search box, click on the submit button, view the results, scroll through the results, and then click on the respective link. Each of these actions can be further decomposed. This approach is intended specifically to help understand the usability of a system for first-time or infrequent users, that is, for users in an exploratory learning mode.

But how does one perform a cognitive walkthrough? The following procedures must be taken into consideration, according to Richardson (2000):

- Define the inputs:
  - Identify the users and tasks, create a description of the interface (screenshot or prototype), and define the action sequences for completing each task.
- Gather the walkthrough team:
  - A facilitator for the discussion and a scribe for recording information.
  - Participants walk through the tasks relating to the user interface.
  - Participants probe the task on hand (e.g. does the label for the correct action match the user's goal? What is the user's goal? etc.).
- Walk through the action sequences for the task(s).
- Record critical information (i.e. the happy (or failure) path(s), problems, etc.).
- Review the interface to fix the problems (i.e. re-implement rapid prototype or new screenshots).
- Repeat (i.e. iterative design, which means prototyping, testing, analyzing, and refining the user interface, and repeating if needed).

4.3.3 GOMS Model

As already mentioned, the GOMS model is a family of techniques for modeling and describing human task performance. However, GOMS models have also been used as a usability inspection method as shown by Card et al. (1983), Lecerof and Paternò (1998), Schrepp (2010), John and Kieras (1996a), and Kieras (2006).

Goals represent the goals that a user is trying to accomplish, usually specified in a hierarchical manner. Operators are the set of atomic-level operations with which a user composes a solution to a goal. Methods represent sequences of operators, grouped together to accomplish a single goal. Selection Rules are used to decide which method to use for achieving a goal when several are applicable.

According to John and Kieras (1996a), the GOMS model has been one of the few extensively recognized theoretical concepts in HCI, and it has been used in real-world design and evaluation situations. Actually, John and Kieras (1996) summarize the previous work on GOMS by offering a unified analysis of GOMS models, showing how these models can be used in design, and describing several examples of the application of GOMS to the design and evaluation of the interfaces for a multiplicity of real-world systems such as the computer-aided design (CAD) system for mechanical design, the space operations database system, the mouse-driven text editor, and so on.

One recent application is the GOMS analysis as a tool to investigate the usability of web units for disabled users (Schrepp, 2010). Guideline compliance is a necessary but not sufficient condition to guarantee the usability of web units by disabled users since efficiency-related issues can be as exclusive for disabled users as violations to basic guidelines. Schrepp explains that GOMS analysis can be adapted to evaluate the efficiency of interface designs for disabled users.

Another application of using GOMS for evaluating user interface is "A GOMS Model for Keyboard Navigation in Web Pages and Web Application" developed by Schrepp and Fischer (2006). Generally speaking, technology should be accessible and usable by users effortlessly, including users with disabilities. According to the study authors, 81% of the websites were incompliant with basic standards for accessibility as recommended by the World Wide Web Consortium* (W3C). Also, the study infers that a task that can be carried out in a minute using a keyboard might need five to ten minutes using a mouse. Many disabled users favor handling desktop and web applications by using the keyboard, as it is often faster than using a mouse. Therefore, offering efficient keyboard support is crucial to increase the usability of those applications. To achieve this, it is necessary to make use of a method to evaluate mouse and keyboard navigation – the GOMS model is a very appropriate method for doing this, as it allows the comparison of diverse methods when dealing with a web application. According to the GOMS model, the average time taken to carry out a task such as following a link will be 11.83 seconds using a mouse and 28.56 seconds using the keyboard. If the time taken to work on a web page using a keyboard is twice or even more than twice the time taken using a mouse, then additional keyboard support should be implemented for this page. Furthermore, the GOMS model can be used to verify whether the level of keyboard support for a web application is appropriate to guarantee that there are no inadmissible drawbacks to keyboard users.

---

* W3C http://www.w3.org/.

4.3.4 Additional User Research and Usability Evaluation Methods

It is worth noting that there are other usability evaluation methods (Law et al., 2008), such as condensed contextual inquiry (CCI), that have been used on a limited scale in industry settings. This section provides an overview of these methods as follows:

■ *Condensed contextual inquiry* (Kantner and Keirnan, 2003): CCI is a context of use method under the user research technique. User Research focuses on understanding user behaviors, needs, and motivations through observation techniques, task analysis, and other feedback methodologies. Traditional CI requires long hours with each user, which usually can take a full day per visit. Although this long session time allows researchers to gather much important information, organizations refrain from spending the time to gather and analyze so much data. Additionally, organizations are reluctant to interrupt employees for such an extended period of time. This method is basically a one-on-one observation of work practice in the user's real context. CCI, in fact, identifies a more restrained suite of concerns to examine than the traditional version of a CI, which often requires a full day per visit. CCI tries to accommodate the restricted time that product development teams have to learn about users' work processes and motivations. The primary difference between traditional CI and CCI is the restricted nature of the work under observation. CCI looks at the bigger picture of the users' motivations and contextual artifacts for accomplishing work. Therefore, this method is not suitable for designing a complex system such as authentication management applications.
■ *Remote usability testing (RUT)*: This method is under the usability testing technique, which is used in user-centered interaction design. RUT is an evaluation method which remotely gathers key data from larger populations (e.g. 40, 80, 120, and so on) in a single usability testing. Participants are observed by usability evaluators in real-time sessions to collect instantaneous behaviors and comments. This method collects performance measures such as the number of errors, types of errors, time on task, and automatic data collection regarding user behavior such as logging (client and server sides), browser logs, etc. According to Law et al. (2008), these requirements have led evaluators to conduct usability testing with larger sample sizes. The RUT method is typically used to compare two or more products, designs, or product features from competitors. According to Paternò and Santoro (2008), the key dimensions for analyzing the different methods for assessing remote usability evaluation are:
  - Type of interaction between the user and the evaluator.
  - Platform employed for the interaction (desktop, mobile, vocal, etc.)

- Techniques used for collecting information about users and their behavior (graphical logs, voice/webcam recordings, eye-tracking, etc.).
- Type of application considered in terms of the implementation environment (Web, Java-based, .NET, etc.).
- The type of the evaluation results provided (task performance, emotional state).

■ *Longitudinal usability evaluations:* This method is also in the context of use method under user research technique. Longitudinal studies follow a user over an extended period of time (i.e. a month or two), with observations made at periodic intervals. After each experiment, for instance, participants are typically asked to answer two questionnaires: one questionnaire intended to assess the hardware, software, and technological issues encountered, and the other one intended to assess the extent and quality of the cooperation between users. According to Sy (2009), the research methods most appropriate for longitudinal studies are:

- Diary studies.
- Usage logs and clickstream/instrumented data analysis.
- Periodic field ethnography.
- Periodic interviewing (both on-site and remote).
- Periodic usability testing (both on-site and remote).
- Retrospectives.

There is no single method that is the most appropriate one for longitudinal studies, but what is critical here is to triangulate data compilation from several methods.

An example is a longitudinal laboratory-based usability evaluation of a healthcare information system (Kjeldskov et al., 2010). The goal of this study was to inquire into the nature of usability problems experienced by novice and expert users and to see to what extent usability problems of a healthcare information system would or would not disappear over time, as the nurses got more familiar with it. The authors conducted a longitudinal study with two main sub-studies: a usability evaluation was conducted with novice users when an electronic patient record system was being employed in a large hospital. After the nurses had used the system in their daily work for fifteen months, the authors repeated the evaluation. The results demonstrate that time does not heal. Even though some problems were not as severe, they still remained after one year of extensive use.

4.4 Usable Security Principles and Guidelines

The research work of Whitten and Tygar (1998 and 1999) on the usability of Pretty Good Privacy (PGP), a public key encryption application, is considered pioneering in the usable security field.

To date, there is no theoretical framework for an inspection method that considers security and usability synergistically for user authentication methods.

However, the HCISec community has been steadily developing research work in usable security guidelines and standards for computer security software *in general (not particularly for user authentication)* such as computer security design principles (Saltzer and Schroeder, 1975), design guidelines for security management systems (Chiasson et al., 2007), guidelines and strategies for secure interaction design (Yee, 2005), design principles and patterns for aligning security and usability (Garfinkel, 2005), and finally, properties of the usability problem for security (Whitten and Tygar, 1998). The next sections describe all of these *security general-purpose* usable security guidelines and standards.

4.4.1 Computer Security Design Principles (Saltzer and Schroeder, 1975)

The work of Saltzer and Schroeder (1975), "The Protection of Information in Computer Systems", presents the basis required for designing and implementing secure software systems. Their principles describe useful practices that are suitable mainly for architecture-level software decisions, regardless of the platform or software language. Software developers, whether they are developing new software or assessing existing software, should always apply these design principles as a benchmark for making their software more secure.

The eight design principles that apply particularly to protection mechanisms are the following:

1. *Keep the design as simple and small as possible*: The most natural way to do any task should also be the most secure way. This well-known principle applies to any aspect of a system, but it deserves emphasis for protection mechanisms for this reason: Design and implementation errors that result in unwanted access paths will not be noticed during normal use (since normal use usually does not include attempts to exercise improper access paths). As a result, techniques such as line-by-line inspection of software and physical examination of hardware that implements protection mechanisms are necessary. For such techniques to be successful, a small and simple design is essential.

2. *Fail-safe defaults*: Access decisions should be based on permission rather than exclusion. This means that the default situation is lack of access, and the protection scheme identifies conditions under which access is permitted. The alternative, in which mechanisms attempt to identify conditions under which access should be refused, presents the wrong psychological base for secure system design. A conservative design must be based on arguments as to why objects should be accessible, rather than why they should not. In a large system, some objects will be inadequately considered, so a default of lack of permission is safer. A design or implementation mistake in a mechanism that gives explicit permission tends to fail by refusing permission, a safe situation, since it will be quickly detected. On the other hand, a design

or implementation mistake in a mechanism that explicitly excludes access tends to fail by allowing access, a failure which may go unnoticed in normal use. This principle applies both to the outward appearance of the protection mechanism and to its underlying implementation.

3. *Complete mediation*: Every access to every object must be checked for authority. This principle, when systematically applied, is the primary underpinning of the protection system. It forces a system-wide view of access control, which in addition to normal operation includes initialization, recovery, shutdown, and maintenance. It implies that a foolproof method of identifying the source of every request must be formulated. It also requires that proposals to improve performance by remembering the result of an authority check be examined skeptically. If a change in authority occurs, such remembered results must be systematically updated.

4. *Open design*: The design should not be secret. The mechanisms should not depend on the ignorance of potential attackers but rather on the possession of specific, more easily protected keys or passwords. This decoupling of protection mechanisms from protection keys permits the mechanisms to be examined by many reviewers without concern that the review may itself compromise the safeguards. In addition, any skeptical user may be allowed to convince him/herself that the system s/he is about to use is adequate for his/her purpose. Finally, it is simply not realistic to attempt to maintain secrecy for any system that receives wide distribution.

5. *Separation of privilege*: Where feasible, a protection mechanism that requires two keys to unlock it is more robust and flexible than one that allows access to the presenter of only a single key. The relevance of this observation to computer systems was pointed out by R. Needham in 1973. The reason for this is that once the mechanism is locked, the two keys can be physically separated, and distinct programs, organizations, or individuals can be made responsible for them. From then on, no single accident, deception, or breach of trust is sufficient to compromise the protected information. This principle is often used in bank safe-deposit boxes. It is also at work in the defense system that fires a nuclear weapon only if two different people both give the correct command. In a computer system, separated keys apply to any situation in which two or more conditions must be met before access is permitted. For example, systems providing user-extendible protected data types usually depend on the separation of privilege for their implementation.

6. *Least privilege*: Every program and every user of the system should operate using the fewest privileges necessary to complete the job. Primarily, this principle limits the damage that can result from an accident or error. It also reduces the number of potential interactions among privileged programs to the minimum for correct operation, so that unintentional, unwanted, or improper uses of privileges are less likely to occur. Thus, if a question arises related to the misuse of a privilege, the number of programs that must be

audited is minimized. Put another way, if a mechanism can provide "fire-walls," the principle of least privilege provides a rationale for where to install the firewalls. The military security rule of "need-to-know" is an example of this principle.

7. *Least common mechanism*: Minimize the number of mechanisms common to more than one user and depended on by all users. Every shared mechanism (especially one involving shared variables) represents a potential information path between users and must be designed with great care to be sure it does not unintentionally compromise security. Furthermore, any mechanism serving all users must be certified to the satisfaction of every user, a job presumably harder than satisfying only one or a few users. For example, given the choice of implementing a new function as a supervisory procedure shared by all users or as a library procedure that can be handled as though it were the user's own, choose the latter option. Then, if one or a few users are not satisfied with the level of certification of the function, they can provide a substitute or not use it at all. Either way, they can avoid being harmed by a mistake in it.

8. *Psychological acceptability*: It is essential that the human interface is designed for ease of use so that users routinely and automatically apply the protection mechanisms correctly. Also, to the extent that the user's mental image of his/her protection goals matches the mechanisms s/he must use, mistakes will be minimized. If s/he must translate his/her image of his/her protection needs into a radically different specification language, s/he will make errors.

4.4.2 Design Guidelines for Security Management Systems (Chiasson et al., 2007)

Although end users are the major concern for the field of usable security, interfaces for security experts are equally important, because the consequences of usability problems can potentially leave entire networks vulnerable to attack. For example, a security analyst, an intrusion detection specialist, or a security administrator (s/he may also take on some of the tasks of a security analyst in smaller organizations) might miss an attack entirely or misdiagnose it. For example, a text interface provides access to fine-grained detail, providing flexible interactions and customizations but it bothers the user with high quantities of data and the need to know the command syntax. On the other hand, a visual interface can provide an overview of the data, which eases the detection of attacks, but it fails to provide fine-grained detail and so some attacks may be missed. Also, despite the fact that the knowledge acquired to design for end users can assist in designing interfaces for security experts, these two user categories (end user and security expert) are quite different in terms of the domain knowledge, their level of responsibility, the amount of information these users have to process, and the consequences of their actions. To this end, Chiasson et al. (2007) developed a preliminary set of design guidelines for security management interfaces as follows:

1. Administrators should reliably and promptly be made aware of the security tasks they must perform.
2. Administrators should be able to figure out how to successfully perform those tasks.
3. Administrators should be able to tell when their task has been completed.
4. Administrators should have sufficient feedback to accurately determine the current state of the system and the consequences of their actions.
5. Administrators should be able to revert to a previous system state if a security decision has unintended consequences.
6. Administrators should be able to form an accurate and meaningful mental model of the system they are protecting.
7. Administrators should be able to easily examine the system from different levels of encapsulation in order to gain an overall perspective and be able to effectively diagnose specific problems.
8. The interface should facilitate the interpretation and diagnosis of potential security threats.
9. Administrators should be able to easily seek advice and take advantage of community knowledge to make security decisions.
10. The interface should encourage administrators to address critical issues in a timely fashion.

These design principles recognize that users will need to make key decisions and be supported in this process. The majority of the interactions will take place because of unpredicted events that the system cannot handle on its own, and as such, it should strive to give clear, pertinent, and enough information so that users are able to precisely identify and address the problem.

Furthermore, it is to be expected that users will occasionally make mistakes when dealing with these novel situations, so the system must allow users to easily revert to a previous state. For example, poor upgrades through security patches can lead to unstable systems that need to be rolled back. Occasional mistakes are unavoidable, and thus the systems must be flexible enough so that recovery is possible.

When faced with a new security threat, it is likely that others are also being similarly attacked. The interface should support and facilitate interaction within the security community not only to more quickly analyze a new threat and determine appropriate countermeasures but also to facilitate propagation of such security measures. Social navigation could also be used to provide trusted feedback about what steps others have taken in similar situations and could be further customized by defining a specific group of trusted sources from which to gather information. Integrating the communication and social navigation into the system could be faster, have less noise, and be harder to spoof than current ad hoc methods. Security systems still generate a sufficiently large number of false alarms to potentially lure administrators into ignoring alarms or deeming them as non-urgent or otherwise lead to situations where it is impossible to address all alarms. This may result in unnecessarily vulnerable systems. The interface should attempt to recognize such situations and encourage the administrators to take corrective action. The interface should alert administrators if the majority of other security professionals have taken some preventative measure that has yet to be addressed in the current system, especially if related to a severe threat given the specific system configuration.

4.4.3 Guidelines and Strategies for Secure Interaction Design (Yee, 2005)

This section presents a preliminary set of guidelines for secure interaction design. The criterion used by Yee (2005) for admitting something as an essential principle is that it should be a *valid* and *non-trivial* concern. Each principle is valid by showing how a violation of the principle would lead to security vulnerability. In the statement of these principles, the term "actor" is used to mean "user or program". The term "authority" only refers to the capability to take a particular action.

1. *Path of least resistance*: The most natural way to do any task should also be the most secure way.
2. *Appropriate boundaries*: The interface should expose, and the system should enforce, distinctions between objects and between actions along boundaries that matter to the user.
3. *Explicit authorization*: A user's authorities must only be provided to other actors as a result of an explicit user action that is understood to imply granting.
4. *Visibility*: The interface should allow the user to easily review any active actors and authority relationships that would affect security-relevant decisions.
5. *Revocability*: The interface should allow the user to easily revoke authorities that the user has granted, wherever revocation is possible.
6. *Expected ability*: The interface must not give the user the impression that it is possible to do something that cannot actually be done.
7. *Trusted path*: The interface must provide an unspoofable and faithful communication channel between the user and any entity trusted to manipulate authorities on the user's behalf.
8. *Identifiability*: The interface should enforce that distinct objects and distinct actions have unspoofably identifiable and distinguishable representations.
9. *Expressiveness*: The interface should provide enough expressive power (a) to describe a safe security policy without unnecessary difficulty and (b) to allow users to express security policies in terms that fit their goals.
10. *Clarity*: The effect of any security-relevant action must be clearly apparent to the user before the action is taken.

4.4.4 Design Principles and Patterns for Aligning Security and Usability (Garfinkel, 2005)

Garfinkel's (2005) doctoral thesis's philosophy was to identify patterns that can make systems that are in fact secure, rather than the conventional goal of creating systems that are in theory securable. This section introduces the six general design principles and patterns for aligning security and usability:

1. *Principle of least surprise*: This principle is an interpretation of Saltzer and Schroeder's (1975) principle of "psychological acceptability". This principle holds that the computer should not surprise the user when the user expects the computer to behave in a manner that is secure. This principle is violated when there is a mismatch between the user's expectations and the computer's implementation.

2. *Principle of good security now*: Computer security is an engineering discipline. Even though it is impossible to have a computer system that is completely secure, there is always a tension between deploying good systems that are available today and waiting for better systems that can be deployed tomorrow. This principle holds that it is a mistake not to deploy good systems that are available now: if good systems are not deployed, end users who are not trained in security will create their own poor security solutions.

3. *Standardized security policies*: Today's security sub-systems provide too many choices and configuration options that are relevant to security. These choices are frequently overwhelming to end users. Worse, relatively minor changes in a security policy or configuration can have a drastic impact on overall security. Most end users need security experts to make decisions for them, because – by definition – end users are not experts. This is not to say that users need to be locked in tightly to a few inflexible policies from which they can never deviate. What is needed is a range of well-vetted, understandable, and teachable policies, and then the ability to make understood, controlled, contained and auditable deviations from these policies when needed.

4. *Consistent meaningful vocabulary*: Usability is promoted when information is presented with a vocabulary that is consistent and meaningful. But there is a natural tendency among computer engineers to be loose with their choice of language. A guiding principle for aligning security and usability is that security information, at least, must be standardized and used consistently.

5. *Consistent controls and placement*: In addition to standardizing vocabulary, it is important that security-related controls be likewise standardized so that similar functionality is presented in a similar manner and in a consistent location in user interfaces.

6. *No external burden*: Security tools must not pose a burden on non-users who do not otherwise benefit from their use. Otherwise, non-users will push back on users through social channels and encourage the users to discontinue the use of the tools.

According to Grudin (1989), these principles should be adapted rationally to the tasks that are at hand since there are many cases in which a simple application of consistent UI rules does not lead to interfaces that are easy to use.

4.4.5 Criteria for Security Software to be Usable (Whitten and Tygar, 1998)

The authors studied the usability of Pretty Good Privacy version 5.0 (a public key encryption application)*, which was considered to have a good GUI, but the results showed that PGP 5.0 was not suitably usable to provide effective security for most users.

Security has some inherent properties that make it a difficult problem domain for user interface design. Design strategies for creating usable security will need to take these properties explicitly into account, and generalized user interface design does not do so. Five main properties are described below.

1. *The unmotivated user property*: Security is usually a secondary goal. People do not generally sit down at their computers wanting to manage their security software; rather, they want to send email, browse web pages, or download software (primary goals), and they want security in place to protect them while they do those things. It is easy for people to put off learning about security, or to optimistically assume that their security is working, while they focus on their primary goals. Designers of user interfaces for security should not assume that users will be motivated to read manuals or look for security controls that are designed to be unobtrusive. Furthermore, if security is too difficult or annoying, users may give up on it altogether.

2. *The abstraction property*: Computer security management often involves security policies, which are systems of abstract rules for deciding whether to grant accesses to resources. The creation and management of such rules is an activity that programmers take for granted but which may be unfamiliar and unintuitive to the wider user population. User interface design for security will need to take this into account.

3. *The lack of feedback property*: The need to prevent dangerous errors is crucial in order to provide good feedback to users but providing good feedback for security management is a difficult problem. The state of a security configuration is usually complex and attempts to summarize it are not adequate. Furthermore, the correct security configuration is the one which does what

---

* PGP (Pretty Good Privacy): A freeware (for non-commercial users) encryption program that uses the public key approach: messages are encrypted using the publicly available key, but the intended recipient can only decipher them via the private key. PGP is perhaps the most widely used encryption program https://www.symantec.com/encryption/.

the user "really wants", and since only the user knows what that is, it is hard for security software to perform much useful error checking.

4. *The barn door property*: The proverb about the futility of locking the barn door after the horse is gone is descriptive of an important property of computer security: Once a secret has been left accidentally unprotected, even for a short time, there is no way to be sure that it has not already been read by an attacker. Because of this, user interface design for security needs to place a very high priority on making sure users understand their security well enough to keep from making potentially high-cost mistakes.

5. *The weakest-link property*: It is well known that the security of a networked computer is only as strong as its weakest component. If a cracker can exploit a single error, the game is up. This means that users need to be guided to attend to all aspects of their security, not left to proceed through random exploration as they might with a word processor or a spreadsheet (application).

4.4.6 Additional Criteria for Security Software to Be Usable (Chiasson et al., 2006)

In their usability study regarding two password managers, Chiasson et al (2006) proposed two additional criteria which actually support the security properties 2 and 3 from Whitten and Tygar (1998) described in Section 4.4.5:

1. *Be able to tell when their task has been completed*: It concerns a usability problem seen in both the Whitten and Tygar (1998) and Chiasson et al. (2006) studies: users were unable to tell whether their task had been successfully completed, and they sometimes incorrectly assumed success. This can cause security vulnerabilities (e.g. as information believed to be secure can be left unprotected).

2. *Have sufficient feedback to accurately determine the current state of the system*: This criterion uses the well-known usability guideline of *feedback*, which is especially important for supporting accurate mental models in security interfaces. Transparency in this case can be dangerous because it leaves users free to make assumptions about the system that could lead to security breaches.

4.4.7 General Security Usability Principles (Identity Management) (Jøsang et al., 2007a)

Direct user involvement in a security service is often required, and a distinction can be made between two types of involvement:

1. A *security action* is when users are required to produce information and security tokens to trigger some security-relevant mechanism. For example, typing and submitting a password is a security action.

2. A *security conclusion*, in turn, is when users observe and assess some security-relevant evidence in order to derive the security state of systems. For example, observing a closed padlock on a browser and concluding that the communication is protected by Secure Socket Layer (SSL) is a security conclusion.

Usability principles related to security actions and security conclusions are described in the following.

- Security action usability principles
    - The users must understand which security actions are required of them.
    - The users must have sufficient knowledge and the practical ability to take the correct security action.
    - The mental and physical load of a security action must be tolerable.
    - The mental and physical load of making repeated security actions for any practical number of transactions must be tolerable.
- Security conclusion usability principles
    - The user must understand the security conclusion that is required for making an informed decision. This means that users must understand what is required of them to support a secure transaction.
    - The system must provide the user with sufficient information for deriving the security conclusion. This means that it must be logically possible to derive the security conclusion from the information provided.
    - The mental load of deriving the security conclusion must be tolerable.
    - The mental load of deriving security conclusions for any practical number of service access instances must be tolerable.

Chapter 5

The Usable Security Protocol Methodology: Define, Identify, and Develop

---

Summary

This chapter starts with an overview of the usable security protocol methodology, its steps, goal, and the logistics behind each step. A discussion is provided on how the methodology puts together the cognitive and computer science approaches for developing a usability inspection method for user authentication. A comprehensive description details the theoretical and demonstrational basis for the verification and validation (V&V) phase (i.e. the process of checking that a software system meets specifications and that it fulfills its intended purpose). Each step of this protocol is explained, and finally, at the end, it is shown how all steps fit together to provide an inspection tool for the design of user authentication methods which is named Usable Security Symmetry (USS).

5.1 Methodology and Architecture

It is worthwhile to mention the definition of a methodology: a system of broad principles or rules from which specific methods or procedures may be derived to interpret or solve different problems within the scope of a particular discipline.

The usable security protocol methodology refers to the framework that is used to structure, plan, develop, and control our usable security symmetry inspection method.

An orderly and sequential seven-step methodology, which makes clear the process of enquiry through which knowledge materializes, is undertaken to generate the protocol architecture as depicted in Figure 5.1. The building blocks start with the primary, secondary, and tertiary data. Then, the cognitive science model (Cognitive Ergonomics) is developed, followed by the computer science model (demonstration) (some tasks might be developed in parallel), each of which forms the theoretical and demonstrational approaches respectively.

The cognitive science and computer science approaches provide the necessary information to develop the following steps and their respective activities:

Step 1: Define the mission and conceptual design objective.

- Formalize a usable security definition.
- Specify task scenario, usability scenario, and security scenario (i.e. use cases).
- Identify users and working contexts.

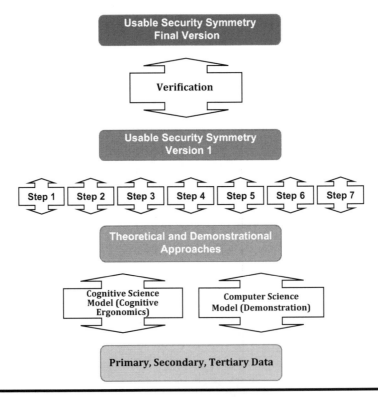

Figure 5.1 The usable security protocol methodology.

Step 2: Identify the most representative user authentication method categories.

- Understand what a user authentication method is, how it works, and what its features are.
- Carry out a classification analysis from the user authentication literature review (i.e. authentication marketplace).
- Perform a comparative analysis of user authentication methods.
- Select the most representative user authentication methods and their categories.

Step 3: Develop the Natural Goals, Methods, Selection Language (NGOMSL).

- Implement the basic operators:
  - Standard primitive external operators.
  - Standard primitive mental operators.
  - Analyst-defined mental operators.
- Total execution time (TET) and total learning time (TLT).
- Calculate total execution time and total learning time for the tasks scenarios:
  - Check business email;
  - Update the SecurID token user interface specification;
  - Make an electronic funds transfer;
  - Access a file on a personal laptop.
- Perform the NGOMSL analysis for each task scenario.

Step 4: Develop the authentication risk-assessment matrix.

- Identify the most critical security vulnerabilities related to web authentication.
- Undertake the risk-assessment matrix.

Step 5: Generate the usable security principles from the NGOMSL model.

- Classify and prioritize the cognitive processes generated by the NGOMSL model.
- Classify and develop a cross-cognitive and usable security principles analysis in order to specify the final usable security principles targeted to user authentication.

Step 6: Formulate the usable security symmetry inspection method.

- Identify the usability factors and usability criteria.
- Define user authentication use cases.
- Specify project lead and development activities.
- Specify usability severity ratings.
- Specify security severity ratings.

Step 7: Demonstrate the usable security symmetry.

■ Present the demonstrational approach through the following:
  – Multifunction Teller Machine (MTM).
  – Virtual private network (VPN) application;
  – RSA SecurID® 700 hardware token authenticator

Next, a first version of the inspection method is released, followed by the V&V phase. Finally, after V&V we culminate with the final inspection method version. As already mentioned, the key point of this protocol is the creation, development, and integration of the inspection method into the requirements and design process of the user authentication method design.

5.2 Define the Mission and the Conceptual Design Objective

5.2.1 Formalize a Usable Security Definition

According to the authors, *usable security* is the study of how security information and usability factors should be handled in both front end and back end, taking into consideration resources and costs. Front end processes are represented by the presentation layer (i.e. interface), and it is regarded as a shared limit through which the information flows (Maffezzini, 2006). The back end, in turn, is represented by the data layer.

5.2.2 Define Task Scenario, Usability Scenario, and Security Scenario

Scenarios describe the stories and context behind why a specific user or user group comes to your site. They note the goals and questions to be achieved and occasionally define the possibilities of how users can achieve them on the site.

The concepts of *task and usability scenarios* have been already fully adopted by the human–computer interaction (HCI) community. However, in the next paragraphs, with regard to the *security scenario*, the authors introduce a novel definition in the context of usable security.

5.2.2.1 Types of Scenarios

Task scenario is a description of the user's tasks to make clear: what the user is trying to achieve (their "practical goal"), the environment within which the user works (task context), and what the user actually does in detail. The purpose of a task scenario is to provide examples of usage as an input to design and to provide a basis for subsequent usability testing. For illustrative purposes, Figure 5.2 shows a typical task scenario notation involving two users in different roles.

TASK SCENARIO 1:	**Check Stock Availability: Part Number and In Store.**
GOAL:	Answer customer query as to current stock holding.
ENVIRONMENT:	Sales office. Personal computer connected to the company database server. Phone with standard headset.
USER:	Direct: Sales Clerk.
	Indirect: Customer.

Customer phones, customer has part number, parts in stock.

SALES CLERK:

> Answer phone.

CUSTOMER:

> "Have you got a 65 Volkswagen Beetle hub-cap, part number GA324 in stock?"

SALES CLERK:

> Indicate manufacturer, Volkswagen.
>
> Indicate model, Beetle.
>
> Indicate part number, GA324.
>
> Receive output stock holding: 13, price: $55.00.
>
> Tell customer: "Yes we have got one. Its price is $55.00."

Figure 5.2   This is an example of a task scenario involving two users in different roles, one a direct and the other an indirect.

Usability scenario is a user scenario which is built around defined user profiles and personas and includes descriptions of common user tasks, as shown in Figure 5.3. The purpose of user scenario development is to define and understand the circumstances in which the user would be likely to use a specific software application. The goal is to build ideal user experiences that will most likely meet the expectation of the user group and result in the successful accomplishment of users' goals.

Security scenario is a description of a task scenario that includes the use of a particular security mechanism. A security scenario can be *tangible* or *intangible*. A *tangible* security scenario includes physical infrastructure, for example, controlling a user's access to a research laboratory using fingerprint (i.e. biometrics), or sending a silent alarm in response to a threat at an MTM. An *intangible* one includes knowledge-based data, for example, a user who enters sensitive information in the registration screen to purchase a concert ticket at an MTM. A security scenario might or might not be a combination of tangible and intangible security scenarios.

More detailed examples of task, usability, and security scenarios are described in following chapters, which describe user authentication use cases for an MTM system such as:

USABILITY SCENARIO: Data Selection and Exploration	
SCENARIO No.:	15
SOURCE:	Users
STYMULUS:	Users are working on a set of data using multiple views.
ENVIRONMENT:	Runtime.
ARTIFACT:	System.
RESPONSE:	The set of available views does not change when switching the operation mode.
USABILITY PATTERNS:	25 - Making Views Accessible
	Users often want to see data from other viewpoints. If certain views become unavailable in certain modes of operation, or if switching between views is cumbersome, the user's ability to gain insight through multiple perspectives will be constrained.

Figure 5.3   Usability scenario example involving data selection.

- Authenticate to an MTM using multipurpose contactless smart card
- Transfer funds to an international bank account
- Buy a concert ticket
- Access your MTM offline with your mobile phone
- Deposit a check by using checking image
- Send a silent alarm.

5.2.3 *Identify Users and Working Contexts*

This section involves identification of the users and their working context. For example, while working in the security domain, the following are some examples of the users who need to be considered.

- *Super administrator.* This IT professional is assigned to administer a mission-critical application. The super admin has all permissions to configure and administrate the system and other administrators, and they are typically the person who would be responsible for planning, deploying, and configuring the software. The super admin can also act as an Approver or Distributor.
- *Security administrator.* The security admin sets up the security aspects of an account, and they are responsible for all the security attributes of a user or role and for the following tasks: assigning and modifying the security attributes of a user, role, or rights profile; creating and modifying rights profiles; assigning rights profiles to a user or role; assigning privileges to a user, role, or rights profile; assigning authorizations to a user, role, or rights profile; removing

privileges from a user, role, or rights profile; and removing authorizations from a user, role, or rights profile.

- *Domain administrator.* The role of the domain admin is to configure and maintain the authentication manager software for the portion of the enterprise for which they are responsible. They can manage, for example, objects such as users, user groups, tokens, and password policy. The domain admin can also act as an Approver or Distributor.
- *Realm administrator.* The realm admin performs all operations in the configured root realm. Authentication properties, authorization policies, data stores, subjects (including a user, a group of users, or a collection of protected resources), and other data can be defined within the realm.
- *System administrator.* The system admin creates user accounts and is responsible for determining who can access the system. The system administrator is responsible for the following tasks: adding and deleting users, adding and deleting roles, and modifying user and role configurations, other than security attributes.
- *Policy administrator.* The policy admin creates and manages policies only. Admin policies are used to regulate the actions that administrators are allowed to do.
- *Helpdesk administrator.* The helpdesk admin is a person who provides first-tier or second-tier helpdesk technical troubleshooting support for end users.
- *Developer.* A developer is a person who designs and writes software.
- *Customer.* A customer is the buyer of the user authentication product. The customer should not be confused with the end user.
- *End User.* The end user is an individual who will ultimately use a software or hardware authenticator to enable him/her to perform a job function (e.g. an end-user uses a hardware token to authenticate to a VPN).

5.3 Identify the Most Representative User Authentication Methods Categories

5.3.1 Understand the User Authentication Method

The first step is to understand the user authentication method, how it works, and what its features are. Details of different authentication methods have already been discussed in Chapter 2.

5.3.2 Carry out a Classification Analysis

The next step in order to identify the most representative user authentication method is to carry out a classification analysis from the user authentication literature review e.g. authentication marketplace.

5.3.3 Comparative Analysis of User Authentication Methods

The next stage is to perform a comparative analysis of user authentication methods. An example has already been presented in Chapter 3 (e.g. see Table 3.3).

5.3.4 Select the Most Representative User Authentication Methods and Their Categories

After the comparative analysis, select the most representative authentication method. These are the tasks scenarios selected for the NGOMSL model:

■ Password/PINs (wired-network-based task): Username and password login operation in a desktop environment.
■ One-time passwords (OTPs) (wireless/token network-based task): Real-time generated OTPs based on the challenge–response method.
■ Out of band authentication (OOBA) (wired and wireless network-based task) (e.g. computer and mobile device such as a smartphone): Utilization of two separate networks working concurrently to authenticate a user (e.g. when a user initiates certain online transactions, the user will be prompted to enter a four-digit code, which is sent via text message to the user's mobile device).
■ Utilization of biometrics (wired network and electronic access control-based task): Logical and physical access control (e.g. a fingerprint).

5.4 Develop the Natural GOMS Language (NGOMSL)

As already mentioned, the NGOMSL model is a method in which learning time and execution time are predicted based on a program-like representation of the procedures that the user must learn and execute to perform tasks with the system. Under NGOMSL, methods are represented in terms of an underlying cognitive theory known as cognitive complexity theory (CCT). This cognitive theory allows NGOMSL to incorporate internal operators (i.e. actions that users execute) such as manipulating working memory information or setting up sub-goals. Therefore, NGOMSL can also be used to estimate the time required to learn how to achieve tasks. To this end, the NGOMSL task analysis identifies and measures the execution and learning times of key perceptual, cognitive, and motor processes undertaken by users. They are based on expert users and well-defined tasks. There is an emphasis on the analysis of those cognitive processes involved in the user authentication processes.

The task scenarios' descriptions and a list of goals, methods, and operators for each user authentication method category are described throughout this current step. All data resources for developing the NGOMSL model have been gathered and/or developed from usability tests and user interviews, authentication token demonstrations, and published research papers for estimating learning and execution times. A vital decision as to what (and what not to) describe was made when developing the NGOMSL analysis, which is that mental processes should be basically treated as "black boxes" due to their overwhelming complexity. This means that trying to explain in detail, for instance, what the reading mental process is would be extraordinarily difficult. Hence, the authors treat users' reading, verification (looking or seeing), forgetting, and thinking mechanisms as "black boxes".

A set of basic operators for constructing the NGOMSL model is described in the next sub-sections.

5.4.1 Classify and Prioritize the Cognitive Processes Generated by the NGOMSL Model

5.4.1.1 Standard Primitive External Operators

The analyst* defines the primitive motor and perceptual operators based on the elementary actions required by the system being analyzed. The standard primitive external operators used in this NGOMSL analysis are the following:

Click mouse button
Move cursor to <target coordinates>
Type <string of characters>

Locate <name> value from screen is equivalent to the process of scanning specific spots on a screen that supply a value for some parameter specified by <name> to determine the location of this value and place the information into working memory. Basically, there should be a Locate operator executed prior to a Double-click or Click, to reflect that before an object can be clicked, its location must be known (e.g. Locate the password field).

Wait for <description>. The waiting time is the time the user is waiting idly for the system's response (e.g. the user has entered his/her username and password, the system processes user authentication, and logs the user into the system).

5.4.1.2 Standard Primitive Mental Operators

As previously mentioned, mental operators are internal actions performed by users (i.e. steps within the methods). These operators include actions like making a basic decision, recalling an item in short-term memory (STM), retrieving information from long-term memory (LTM), etc. The mental operators used in this NGOMSL analysis are described in the following:

---

* The "Analyst" is the person who performs a GOMS analysis as referred by (Kieras, 1996).

■ Read <name> value from screen is equivalent to the process of interpreting characters on a screen that supply a value for some parameter specified by <name> and placing the information into working memory (e.g. read the password displayed on the digital readout window on the SecurID token).

■ Verify <name> value from screen is equivalent to the process of representing how the user is expected to notice and make use of that feedback information. A verify operator should normally be included at the point where the user must commit to the entry of information (e.g. verify that the password has been correctly typed in).

For memory storage and retrieval, the memory operators reflect the distinction between LTM and STM as they are typically used in computer operation tasks. The standard primitive mental operators used in this analysis are the following:

■ Recall <LTM-object-description>: Recall means to fetch from LTM. Searches LTM for an item whose specified properties have the specified values and stores its symbolic name in STM.

■ Recall <WM-object-description>: Recall means to fetch from STM. Searches STM for an item whose specified properties have the specified values and stores its symbolic name in STM.

■ Retrieve <WM-object-description>: Retrieve means to get back an item from memory, which can either be from LTM or STM, during the method execution.

■ Retain <WM-object-description>: Retain means to store in working memory.

■ Forget <WM-object-description>: Forget means that the information is no longer needed and, thus, can be deliberately dropped from working memory.

■ Listen <Auditory stimulus-object-description>: Listen (auditory stimulus) means that the user listens to either speech or sound inputs. After a standard time delay representing auditory working memory decay time (currently 1000 ms), this object is deleted and the auditory stimulus information is no longer available.

■ Return with goal accomplished is a basic flow of control. A sub-method is invoked by asserting that its goal should be accomplished, and it returns here when the goal has been accomplished. The operator Return with goal accomplished is analogous to an ordinary Return statement, and it marks the end of a method.

5.4.1.3 Analyst-Defined Mental Operators

Analyst-defined mental operators represent psychological processes that are too complex to be practical to designate as methods in the GOMS model. The designer can in fact circumvent these processes by defining operators that act as placeholders for the mental activities as follows:

■ Think-of <description> represents a process of thinking of a value for some parameter designated by <description> and putting the information into working memory (Kieras, 1996) (e.g. Think-of a VPN client padlock icon which indicates a secure connection).
■ Read <name> value from screen is equivalent to the process of interpreting characters on a screen that supply a value for some parameter specified by <name> and placing the information into working memory (e.g. read the password displayed on the digital readout window on the SecurID token).

As stressed by Kieras (1996), the methods have been represented at the standard primitive operator level, so the calculations for predicting learning and execution times will generate realistic, accurate, and useful results.

5.4.2 Understand the Total Execution Time and Total Learning Time

NGOMSL predicts the learning time that users will take to learn the procedures represented in the GOMS model and the execution time users will take to execute particular task instances by following those procedures. It is important here to understand the difference between these two usability measures.

5.4.2.1 The Total Execution Time

The TET comprises the methods, steps, and operators needed to carry out a specific task. The time needed to complete a task instance is determined by the number and content of NGOMSL statements that have to be executed in order to get that particular task done. The time needed by each statement is the sum of a small fixed time for the statement plus the time required by any external or mental operator executed in the statement. For example, each NGOMSL statement (i.e. each step) is assumed to require a small fixed time to execute, and any operators in the statement, such as a keystroke, will then take additional time depending on the operator.

Execution Time = NGOMSL statement time

+Primitive External Operator Time

+Waiting Time

Execution Time = Time for the execution of the methods required to perform the task by the user.

NGOMSL statement time = Number of statements executed multiplied by 0.1 seconds.

Primitive External Operator Time = Total of times for primitive external operators.

Waiting Time = Total time when user is inactive while waiting for the system's response. The authors highlight that the waiting time is irrelevant due to the fact that the system is considered to be fast enough, and additionally it is not the main focus of our GOMS analysis related to user authentication. Therefore, it will be not measured and indicated in the time measurement analysis.

5.4.2.2 The Total Learning Time

The TLT is the total number and length of all methods. The time to learn a set of methods is fundamentally specified by the total length of the methods, which is provided by the number of NGOMSL statements in the GOMS model for the interface. This is the quantity of procedural knowledge that the user has to acquire in order to know how to use the system for all tasks under consideration (Kieras, 2006). The time required to learn how to perform the methods themselves is defined as the Pure Learning Time (PLT).

> **Total Learning Time = Pure Method Learning Time**
> **+LTM Item Learning Time**   .
> **+Training Procedure Execution Time**

Total Learning Time = the total time needed to complete a training process.
Pure Method Learning Time = Learning Time Parameter multiplied by the number of NGOMSL statements to be learned (i.e. the time required to learn how to perform the methods). Learning Time Parameter = 30 seconds for rigorous procedure training or 17 seconds for a typical learning situation. Retrieving a chunk from memory: A chunk is a common unit such as a file name, command name, or abbreviation. For instance, if the user wants to list the contents of directory foo, they need to retrieve two chunks, dir and foo, each of which takes a mental (M) operator. According to (Kieras, 1996), there is no recognized and verified method for counting how many chunks are implicated in "to-be-memorized" information, so what is presented next is heuristic-based information. Count the number of chunks as follows:

- One chunk for each common pattern in the retrieval cue.
- One chunk for each common pattern in the retrieved information.
- One chunk for the association between the retrieval cue and the retrieved information.

For instance, presume that the "to-be-stored" association for a command is *move cursor right by a word is CTRL-RIGHT ARROW*. Then:

(move cursor right) (by a word) = Two chunks for retrieval cue
(ctrl) (right-arrow) = Two chunks for retrieved information
Association between the two = One chunk

Long-Term Memory (LTM) Item Learning Time = the time required to memorize items that will be retrieved from LTM during method execution. It estimated that in general this takes approximately 6 seconds multiplied by the number of LTM chunks (i.e., a common unit such as a file name, command name, or abbreviation) to be learned (Gong, 1993).

Training Procedure Execution Time = If the procedures to be used in training are known, it may be useful to estimate the total learning time by adding the time required to execute the training procedures. The following keystroke-level model (KLM) GOMS operators have been used for some of the operators' duration times (Kieras, 2001) according to the following table. Although those KLM operators have been counted and included as part of the time calculations, they are not explicitly shown in **Sub-Steps 1, 2, 3, and 4** in the next sections as specific values within the time measurements so as to simplify data presentation. When considering a user's typing skills, it is important to note that the typing time depends on the users' typing skill which has been specified by the average skilled typist (55 wpm, i.e., wpm means words per minute and it is a measure of input or output speed) = 0.20 secs (Card et al., 1983).

Operator	Abbreviation	Duration (secs)
Mental	M (1)	1.20
Keystroke <key>	K	0.28
MouseDown or Up	B	0.10
Click (mouseDown & Up)	BB	0.20
Homing (2)	H	0.40
Double-Click	BBBB	0.40
Point w/mouse	P (3)	1.10
Type characters	T(n) (4)	0.20
System Response	R(5)	t

(1) Thinking time; (2) Homing is the process of determining the location of something, and going to it; (3) Point with mouse to a target on a display; (4) T(n): Type a sequence of N characters on a keyboard; (5) The system response time during which the user has to wait for the system. The duration (t) can drastically vary depending on the system being analyzed.

5.4.2.3 Example of TET and TLT

An example of TET and TLT time calculations for Access a file on a personal laptop, Method for goal: Log into the system, follows:

Total Execution Time (TET):

Method for Goal: Log into the System	NGOMSL Statement (secs)	Operator (Type)	Operator Time (secs) (1)	Sub-Total Execution Time (secs)
Step 1. Read fingerprint logon Welcome screen containing finger image and "Password" field on the laptop computer.	0.10	M	2.01	2.11
Step 2. Refer to the Universal Serial Bus (USB)-based biometric fingerprint reader.	0.10	H	2.01	2.11
Step 3. Locate the fingerprint sensor on the USB fingerprint reader.	0.10	M	2.01	2.11
Step 4. Move finger to the USB fingerprint reader.	0.10	P	1.01	1.11
Step 5. Position last knuckle joint over the center of the fingerprint sensor.	0.10	H	1.01	1.11
Step 6. Swipe the finger without lifting it over the fingerprint sensor.	0.10	P	1.21	1.31
Step 7. Verify that you have been granted access to the system.	0.10	M	1.11	1.21
Total Execution Time (secs)	**9.16**			

(1) A 0.01 milliseconds have been added to the Operator Time as a margin of error.

Total Learning Time (TLT):

Total Learning Time = Pure Method Learning Time

+Long - Term Memory Item Learning Tin

+Training Procedure Execution Time

Total Learning Time = 119secs + 30 secs + 0 secs = 149 secs

$\big($Pure Method Learning Time = Learning Time Parameter

×Number of NGOMSL Statements to be learned

->17 secs × 7 steps = 119 secs$\big)$

The following task scenarios with their corresponding authentication methods have been created to develop the NGOMSL model:

- *Check business email* incorporates the username and password login authentication method;
- *Update the SecurID token user interface specification* incorporates OTPs;
- *Make an electronic funds transfer* incorporates OOBA;
- *Access a file on a personal laptop* incorporates biometrics (fingerprint recognition).

These tasks were conceived only to serve as a basis for demonstrating their authentication aspects. Security is a secondary goal for many users, an indispensable step in the way of achieving their primary goals such as the task scenarios mentioned previously. Therefore, it would be odd to describe only the authentication activity given that users don't authenticate to a system and do nothing. There is always a goal involved when authenticating to a system.

As already mentioned, a method is a series of steps that accomplishes a goal. A step in a method typically consists of an external operator, such a pressing a key, or a set of mental operators involved in setting up and accomplishing a sub-goal. Much of the work in analyzing a user interface consists of specifying the actual steps that users carry out in order to accomplish goals, so describing the methods is the focus of the task analysis. According to NGOMSL, the structure for a method is as follows:

Method for goal: <goal description>
Step 1. <operator>...
Step 2. <operator>...
Step 3. <operator>...

...

Step n. Return with goal accomplished.

The NGOMSL analysis lays down the foundation for the design of the usable security inspection method. Finally, this section presents the results of quantitative and qualitative aspects of NGOMSL applied to user authentication.

5.4.3 Calculate Total Execution Time and Total Learning Time for Tasks Scenarios

5.4.3.1 TASK: Check Business Email

TASK_SCENARIO: T1	Check business email
TASK CATEGORY	Wired-network-based task
USER AUTHENTICATION METHOD	Username and Password Login

5.4.3.1.1 Generate Task Description

Password authentication is the most common method of authentication, and also one of the least secure. Basically, the computer asks the user to type in a username and a password. The computer searches the system's password file for an entry matching the username in the database. If the password in that entry matches the password just typed, then the login succeeds. This task scenario makes use of the Microsoft Windows NT* operating system, which controls user access to systems within and across domains (i.e. local and remote access). When a user logs on to an NT system, NT validates the user's account and authorizes access to the appropriate system or domain (i.e. Windows NT authentication).

5.4.3.1.2 Describe a List of High-Level User Goals

The user's topmost goal is: Check business email. The set of a user's high-level goals includes the following:

- Log into the system
- Open Microsoft Office Outlook (MOO)
- Read email Message
- Return with goal accomplished.

This analysis is based on the premise that the main topic of this book is about user authentication, so a particular level of analysis and granularity is required for the "Log into the system" high-level operator. This operator is therefore decomposed into finer levels (i.e. a series of lower-level, or primitive, operators), whereas

---

* Microsoft Windows NT <https://en.wikipedia.org/wiki/Windows_NT>.

the "Open Microsoft Office Outlook" and "Read email message" high-level operators function as supportive methods within the task and are therefore not decomposed. This is in fact a common decision-making point when undertaking GOMS:
specifically, whether the operator should or should not be decomposed into finer
levels depending on the finest grain level of analysis desired by the analyst. It is
worth noting that users do not authenticate to a system per se since *authentication
is not a goal but rather a means* of accomplishing a goal within the context of this
task: check business email.

■ **Define operators and write methods for accomplishing user goals**

- **Method for goal: Log into the system**
  • Step 1. Read Windows Logon Welcome screen on the desktop
    computer.
  • Step 2. Locate Ctrl+Alt+Del key combination on the keyboard and
    simultaneously hold them down.
  • Step 3. Verify that the Windows pop-up window is opened.
  • Step 4. Locate the username field on the screen.
  • Step 5. Move the cursor to the username field.
  • Step 6. Recall the username "jdoe", retrieve it from LTM, and retain it.
  • Step 7. Type the username in the username field.
  • Step 8. Verify that the username has been correctly typed in.
  • Step 9. Forget the username.
  • Step 10. Locate the password field.
  • Step 11. Move the cursor to the password field.
  • Step 12. Recall that the password is "Boat6paper!", retrieve it from
    LTM, and retain it.
  • Step 13. Type the password in the password field.
  • Step 14. Verify that the password has been correctly typed in.
  • Step 15. Forget the password.
  • Step 16. Locate the "Submit" button.
  • Step 17. Double-click the "Submit" button.
  • Step 18. Verify that you have been granted *access* to the system.
  • Step 19. Return with the goal accomplished.
- **Method for goal: Open Microsoft Office Outlook**
  • Step 1. Locate the Microsoft Office Outlook icon in the task bar.
  • Step 2. Double-click on the Microsoft Office Outlook icon.
  • Step 3. Verify that Microsoft Office Outlook has opened up the
    inbox.
  • Step 4. Return with the goal accomplished.
- **Method for goal: Read email message**
  • Step 1. Double-click on any email message row.
  • Step 2. Verify that the email message has been opened.
  • Step 3. Return with the goal accomplished.

5.4.3.1.3 Estimate Total Execution Time

The time measurement related to the execution time for **TASK_SCENARIO: T1 Check business email** is listed in seconds in the following.

Method for Goal: Log into the System	Sub-Execution Time(s)
Step 1. Read Windows Logon Welcome screen on the desktop computer.	1.21
Step 2. Locate Ctrl, Alt, and Del keys on the keyboard.	1.21
Step 3. Hold down Ctrl+Alt+Del keys simultaneously.	0.77
Step 4. Verify that the Windows pop-up window is opened.	1.21
Step 5. Locate the username field on the screen.	1.21
Step 6. Move the cursor to the username field.	1.11
Step 7. Recall username "jdoe", retrieve it from LTM, and retain it.	1.21
Step 8. Type the username in the username field.	2.16
Step 9. Verify that username has been correctly typed in.	1.21
Step 10. Forget the username.	1.21
Step 11. Locate the password field.	1.21
Step 12. Move the cursor to the password field.	1.11
Step 13. Recall that the password is "Boat6paper!" retrieve it from LTM,& retain it.	1.21
Step 14. Type the password in the password field.	2.16
Step 15. Verify that password has been correctly typed in.	1.21
Step 16. Forget the password.	1.21
Step 17. Locate the "Submit" button.	1.21
Step 18. Double-click on the "Submit" button.	0.21
Step 19. Verify that you have been granted *access* to the system.	1.21
Total Execution Time	23.25

Method for Goal: Open the Microsoft Office Outlook	Sub-Execution Time(s)
Step 1. Locate the MOO icon in the task bar.	1.21
Step 2. Double-click on the MOO icon.	0.41
Step 3. Verify that MOO has opened up the inbox.	1.21
Total Execution Time	2.83

Method for Goal: Read email Message	Execution Time(s)
Step 1. Double-click on any email message row.	0.41
Step 2. Verify that the email message has been opened.	1.21
Total Execution Time	1.62

The Total Execution Time for **TASK_SCENARIO: T1 Check business email** is shown in the following:

TASK_SCENARIO: T1 Check Business email	
Methods	*Sub-Execution Time(s)*
Method for goal: Log into the system	23.25
Method for goal: Open Microsoft Office Outlook	2.83
Method for goal: Read email message	1.62
Total Execution Time	27.70

5.4.3.1.4 Estimate Total Learning Time

The total learning time for **TASK_SCENARIO: T1 Check business email** is shown in the following:

Total Learning Time = Pure Method Learning Time

+ LTM Learning Time

+Training Procedure Execution Time

Total Learning Time = 408 secs + 24 secs + 0 secs = 432 secs

5.4.3.2 TASK: Update the SecurID Token
User Interface Specification

TASK_SCENARIO: T2	Update the SecurID token user interface specification
TASK CATEGORY	Hardware token and wired-network-based task with real-time-generated OTPs based on the challenge–response authentication method.
USER AUTHENTICATION METHOD	One-time passwords

5.4.3.2.1 Generate Task Description

The typical scenario where OTP is used is in conjunction with Secure Socket Layer (SSL)*/virtual private network[†] servers and web portals.

In a simple scenario, John Doe, a user, connects to the VPN application and opens it in his computer. John enters his username (jdoe) and a four-digit numeric PIN (7234) in the username and passcode fields in the login application screen. Then John types and appends the six-digit numeric token number (currently displayed on his hardware token, e.g. 435961) to the entered PIN on the login screen. This whole number (7234435961) becomes the **passcode,** which is in fact the combination of the PIN and the token code. John clicks "OK". If the authentication is successful, the authentication server validates the passcode and grants the user access to the network.

Pre-condition for this task scenario: The authentication server randomly generates OTPs (six-digit numeric token numbers) on the user token display (e.g. 435961). John has possession of a hardware authentication token which has been assigned to him by the company's IT administrator.

5.4.3.2.2 Describe a List of High-Level User Goals

The topmost goal for users is **Update the SecurID token user interface specification** using a version control and content collaboration software[‡]. The set of high-level user goals considered are:

■ Open the EMC virtual private network application.
■ Log into the system.

---

* Secure Socket Layer (SSL) is a protocol developed by Netscape for transmitting private documents via the Internet. SSL uses a cryptographic system that uses two keys to encrypt data – a public key known to everyone and a private or secret key known only to the recipient of the message.
† Virtual Private Network (VPN) is a private network that uses a public network (usually the Internet) to connect remote sites or users together.
‡ https://www.perforce.com/products/helix-core.

- Get authorization from the system to the protected resource.
- Enable the version control and content collaboration system.
- Open the SecurID token user interface specification.
- Return with goal accomplished.

5.4.3.2.3 Define Operators and Write Methods for Accomplishing User Goals

- **Method for goal: Open the EMC VPN application (EMC, 2006)**
  Step 1. Locate the EMC VPN icon in the bottom taskbar on the screen in Windows.
  Step 2. Move mouse over the EMC VPN icon and double-click the icon of the VPN.
  Step 3. Verify that the EMC VPN pop-up window is opened.
  Step 4. Return with the goal accomplished.
- **Method for goal: Log into the system**
  Step 1. Locate and verify that the "America's East Coast" connection entry is highlighted in the EMC VPN pop-up window.
  Step 2. Move the mouse over to it.
  Step 3. Double-click it with left mouse button.
  Step 4. Verify that the status bar on the bottom left corner of the pop-up window is displaying "Authenticating user...".
  Step 5. Verify that the "VPN Client | User Authentication for America's East Coast" secondary login pop-up window is opened.
  Step 6. Verify that the username field has been automatically filled in (e.g. joedoe).
  Step 7. Verify that the cursor is automatically placed within the "Passcode" field.
  Step 8. Recall the four-digit personal identification number (PIN), retrieve it from LTM, and retain it.
  Step 9. Type the four-digit PIN in the "Passcode" field.
  Step 10. Verify that asterisks are displayed while entering the PIN within the "Passcode" field.
  Step 11. Forget PIN.
  Step 12. Refer to the SecurID 700 token to get the ever-changing (i.e. every 30 seconds) six-digit numerical password.
  Step 13. Read the six-digit numerical password displayed on the digital read-out window on the SecurID token.
  Step 14. Retain, memorize, and store the six-digit numerical password in the STM.
  Step 15. Retrieve the six-digit numerical password from the STM.
  Step 16. Verify that the cursor is in the correct place within the "Passcode" field.

Step 17. Append the six-digit numerical password to the PIN that has been already entered in the "Passcode" field.

Step 18. Verify that asterisks are displayed while entering the six-digit numerical password in the "Passcode" field.

Step 19. Forget the six-digit numerical password.

Step 20. Move the mouse over to the "OK" button.

Step 21. Double-click the "OK" button.

▪ **Method for goal: Get authorization from the system to the protected resource.**

Step 1. Wait for the authentication server to the check user's username and passcode against the database.

Step 2. Verify that the system displays "Contacting the security gateway at 137.69.115.17..." message in the status bar at the bottom left corner, and ensure that the progress bar at the bottom right is running (i.e., System displays the Internet Protocol (IP) address of the authentication server).

Step 3. Verify that the system displays the "Negotiating security policies..." message (after successfully contacting the security gateway) in the status bar at the bottom left corner, and ensure that the progress bar at the bottom right is running.

Step 4. Verify that the system displays the "Connected to America's East Coast" message (after successfully negotiating the security policies) in the status bar at the bottom left corner, and ensure that the progress bar at the bottom right is running.

▪ **Method for goal: Enable the version control and content collaboration system.**

Step 1. Double-click on the "Continue" button to enable the protected network resource.

Step 2. Verify that the system displays the "VPN Dialer | Banner" pop-up window.

Step 3. Read the "VPN Dialer | Banner" statement.

Step 4. Locate the "VPN Dialer | Banner" padlock icon, think of it as locked (i.e. user is *connected* with the version control and content collaboration), in the status bar bottom right corner, and put this information into the STM.

Step 5. Think also of the "VPN Dialer | Banner" padlock icon as indicating that it is a fully secure connection.

Step 6. Return with goal accomplished.

▪ **Method for goal: Open the SecurID token user interface specification.**

Step 1. Locate the SecurID token user interface specification in the directory of the version control and content collaboration.

Step 2. Double-click with left mouse button.

Step 3. Go to the "Request a Token" section in the specification.

Step 4. Add a new software token type.

Step 5. Return with goal accomplished.

5.4.3.2.4 Estimate Total Execution Time

The time measurement related to the *execution time* for **TASK_SCENARIO: T2 Update the SecurID token user interface specification** is listed in seconds in the following.

Method for Goal: Open the EMC VPN Application	Sub-Execution Time(s)
Step 1. Locate the EMC VPN icon in the bottom taskbar on the screen in Windows.	1.21
Step 2. Move mouse over the EMC VPN icon and double-click the icon of the EMC VPN.	1.11
Step 3. Verify that the EMC VPN pop-up window is opened.	1.21
Total Execution Time	3.53

Method for Goal: Log into the System	Sub-Execution Time(s)
Step 1. Locate and verify that the "America's East Coast" connection entry is highlighted in the EMC VPN pop-up window.	1.21
Step 2. Move mouse over to it.	1.11
Step 3. Double-click it with left mouse button.	0.41
Step 4. Verify that the status bar on the bottom left corner of the pop-up window is displaying "Authenticating user..."	1.21
Step 5. Verify that the "VPN Client	User Authentication for America's East Coast" secondary login pop-up window is opened.	1.21
Step 6. Verify that username field has been automatically filled in (e.g. joedoe).	1.21
Step 7. Verify that the cursor is automatically placed within the "Passcode" field.	1.21
Step 8. Recall the four-digit personal identification number (PIN), retrieve it from LTM, and retain it.	1.51
Step 9. Type the four-digit PIN within the "Passcode" field.	2.16
Step 10. Verify that asterisks are displayed while entering the PIN within the "Passcode" field.	1.21

Step 11. Forget PIN.	1.21
Step 12. Refer to the SecurID 700 token to get the ever-changing (i.e. every 30 seconds) six-digit number password.	2.1
Step 13. Read the six-digit number password displayed on the digital readout window on the SecurID token.	3.0
Step 14. Retain, memorize, and store the six-digit number password in the STM.	1.51
Step 15. Retrieve the six-digit number password from STM.	1.21
Step 16. Verify that the cursor is at the correct place in the "Passcode" field.	1.21
Step 17. Append the six-digit number password to the PIN that has been already entered in the "Passcode" field.	1.50
Step 18. Verify that asterisks are displayed while entering the six-digit number password in the "Passcode" field.	1.21
Step 19. Forget the six-digit number password.	1.21
Step 20. Move mouse over to "OK" button.	1.11
Step 21. Double-click the "OK" button.	0.41
Total Execution Time	28.13

Method for Goal: Get Authorization from the System to the Protected Resource	Sub-Execution Time(s)
Step 1. Wait for the authentication server to check user's username and passcode against the database.	1.21
Step 2. Verify that the system displays the "Contacting the security gateway at 137.69.115.17..." message in the status bar bottom left corner, and ensure that the progress bar at the bottom right is running.	1.21
Step 3. Verify that the system displays the "Negotiating security policies..." message (after successfully contacting the security gateway), in the status bar at the bottom left corner, and ensure that the progress bar at the bottom right is running.	1.21

Step 4. Verify that the system displays "Connected to America's East Coast" message (after successfully negotiating the security policies) in the status bar at the bottom left corner, and ensure that the progress bar bottom right is running.	1.21
Total Execution Time	4.84

Method for Goal: Enable the Version Control and Content Collaboration System	Sub-Execution Time(s)
Step 1. Double-click on "Continue" button to enable the protected network resource.	0.41
Step 2. Verify that the system displays the "VPN Dialer | Banner" pop-up window.	1.21
Step 3. Read the "VPN Dialer | Banner" statement.	1.11
Step 4. Locate the "VPN Dialer | Banner" padlock icon, think of it as locked (i.e. user is *connected* with the version control and content collaboration system) in the status bar at the bottom right corner, and put this information into the STM.	1.51
Step 5. Think also of the "VPN Dialer | Banner" padlock icon as indicating that it is a fully secure connection.	0.29
Total Execution Time	4.53

Method for Goal: Open the SecurID Token User Interface Specification	Sub-Execution Time(s)
Step 1. Locate the SecurID token user interface specification in the directory of the version control and content collaboration system.	1.21
Step 2. Double-click with left mouse button.	0.41
Step 3. Go to the "Request a Token" section in the specification.	2.21
Step 4. Replace token type wording from RSA SecurID Toolbar to RSA SecurID Windows Mobile[a]	0.45
Total Execution Time	4.28

[a] RSA Security LLC. September 1st, 2018 <https://community.rsa.com/community/products/securid/software-token-windows-phone>

The Total Execution Time for **TASK_SCENARIO: T2 Update the SecurID token user interface specification** is shown in the following:

TASK_SCENARIO: T2 Update the SecurID Token User Interface Specification	
Methods	*Sub-Execution Time(s)*
Method for goal: Open the EMC VPN application	3.53
Method for goal: Log into the system	28.13
Method for goal: Get authorization from the system to the protected resource.	4.84
Method for goal: Enable the version control and content collaboration software.	4.53
Method for goal: Open the SecurID token user interface specification.	4.28
Total Execution Time	45.31

5.4.3.2.5 Estimate Total Learning Time

The total learning time for **TASK_SCENARIO: T2 Update the SecurID token user interface specification** is shown in the following:

Total Learning Time = Pure Method Learning Time

+LTM Learning Time

+Training Procedure Execution Time

Total Learning Time = 629 secs + 42 secs + 0 secs = 671 secs

5.4.3.3 *TASK: Make an Electronic Funds Transfer*

TASK_SCENARIO: T3	Make an electronic funds transfer
TASK CATEGORY	Wired- and wireless-network-based task
USER AUTHENTICATION METHOD	Out of band authentication

5.4.3.3.1 Generate Task Description

Out of band authentication is essentially the use of two separate networks, for instance, wired and wireless networks, working concurrently in order to authenticate a user. Consider this scenario: Alice, who has her iPhone mobile device number stored on the bank authentication server, wants to transfer an amount of money to a different bank account. First, she logs onto the bank's website with her credentials (username and password) on her desktop computer. Second, she goes to the money transfer section and selects from the drop-down menu list to transfer more than $15,000 to another bank account; this selection triggers the server to send a code to Alice because the bank's security policy states that this type of transaction requires an additional authentication method for security purposes. Third, the bank sends a code to Alice's iPhone via Short Messaging Service (SMS). Finally, Alice types this code in the text field on the bank's website screen in her desktop and clicks "Submit"; if this code matches the one the bank has just sent, then the transaction is successful.

5.4.3.3.2 Describe a List of High-Level User Goals

The topmost goal of the use is: **Transfer $15,000 to the Bank of America.** The set of the user's high-level goals considered are:

- Go to the bank website.
- Log on to the system.
- Make an electronic funds transfer.
- Return with goal accomplished.

5.4.3.3.3 Define Operators and Write Methods for Accomplishing User Goals

- **Method for goal: Go to the bank website.**
  Step 1. Open the web browser on a desktop computer.
  Step 2. Type the bank's website address.
  Step 3. Return with the goal accomplished.
- **Method for goal: Log on to the system.**
  Step 1. Locate the "Online Banking" logon section on the bank's home page.
  Step 2. Read your online ID that has been already filled in by the system (e.g. a saved online ID such as "go4t****").
  Step 3. Locate the "Sign In" button.
  Step 4. Move the cursor to the "Sign In" button.
  Step 5. Click on the "Sign In" button with the left mouse button.
  Step 6. Verify that it is a secure connection by checking if the bank's address bar contains the prefix "https" when the "Confirm that your SiteKey is correct" page is displayed.
  Step 7. Verify that it is a secure connection by checking that the Uniform Resource Locator (URL) is correct and no errors were encountered on the same page.
  Step 8. Think of the closed yellow padlock icon the bottom right of the Windows task bar as indicating that it is a secure connection.
  Step 7. Locate the SiteKey phrase field.
  Step 8. Read the SiteKey phrase.

Step 9. Recognize the SiteKey phrase (e.g. "Whales are fascinating creatures"), retrieve it from LTM, and retain it.

Step 10. Verify that the SiteKey phrase is correct according to the user's online account setup.

Step 11. Read the SiteKey image.

Step 12. Recognize your SiteKey image (e.g. "a whale image"), retrieve it from LTM, and retain it.

Step 13. Verify that the SiteKey image (e.g. "a whale image") is correct according to the user's online account setup.

Step 14. Locate the "Passcode" field.

Step 15. Verify that the cursor has been already placed in the "Passcode" field on the page load.

Step 16. Recall the passcode (e.g. Light7ocean), retrieve it from LTM, and retain it.

Step 17. Type the passcode within the "Passcode" field.

Step 18. Verify that asterisks are displayed while entering the passcode within the "Passcode" field.

Step 19. Verify that the passcode has been correctly typed in.

Step 20. Forget passcode

Step 21. Locate the "Sign In" button.

Step 22. Click on "Sign In" button with left mouse button

Step 23. Verify that the system has been granted access and the "Accounts" page is displayed.

■ **Method for goal: Make an electronic funds transfer.**

Step 1. Verify that you are in the "Accounts Overview" tab.

Step 2. Locate the "Transfers" tab and click on it.

Step 3. Verify that you are in the "Transfers" page.

Step 4. Verify that you are in the "Make Transfer" sub-tab.

Step 5. Locate the "From" field and select the account.

Step 6. Locate the "To" field and select the account.

Step 7. Locate the "Amount" field and type the amount of $15,000.

Step 8. Locate the "Continue" button and click on it.

Step 9. Read a warning message on the top of the next page which says that "You will have to provide an additional authentication credential in order to proceed with transfers over $10,000. Do you want to continue?"

Step 10. Verify the "OK" and "Cancel" buttons.

Step 11. Locate the "OK" button and click on it.

Step 12. Verify that you are in the "Additional Authentication Credentials" page.

Step 13. Locate the instructions section on this page.

Step 14. Read the instructions about the four-digit numerical code that has to be sent to the iPhone mobile device using SMS. After 15 seconds, verify that an SMS text message has been sent by the bank to the iPhone mobile device. Then enter this same code in the "Code" field on this page.

Step 15. Wait for a sound alert for SMS notification to the iPhone after 15 seconds, indicating that the four-digit numerical code has been sent by the bank. Then open your "Messages" application on your iPhone.

Step 16. Hear the auditory stimulus in the form of a sound alert for SMS notification with the four-digit numerical code that has been sent by the bank to the iPhone, retrieve it from the auditory LTM, and retain it.

Step 17. The auditory stimulus information (sound alert) disappears from LTM and is no longer available.

Step 18. Open your "Messages" application.

Step 19. Verify that you have received an SMS text message with the four-digit numerical code sent by the bank.

Step 20. Read the four-digit numerical code directly from the subject field of the message row.

Step 21. Memorize the four-digit numerical code and retain it in the STM.

Step 22. Locate the "Code" field on the "Additional Authentication Credentials" web page.

Step 23. Locate the "Code" field on the page.

Step 24. Type the four-digit numerical code in the "Code" field.

Step 25. Verify that asterisks are displayed while entering the code within the "Code" field.

Step 26. Forget code.

Step 27. Verify the "Send" and "Cancel" buttons are displayed below the "Code" field on the page.

Step 28. Click on the "Send" button with the left mouse button.

Step 29. Verify that you have been directed to the confirmation page which states that you have successfully transferred the amount of $15,000 to the desired destination account.

Step 30. Return with the goal accomplished.

5.4.3.3.4 Estimate Total Execution Time

The time measurement related to the *execution time* for **TASK_SCENARIO: T3 Transfer $15,000 to the Bank of America** is listed in seconds below:

Method for Goal: Go to the Bank Website	Sub-Execution Time(s)
Step 1. Open the web browser in a desktop computer.	0.91
Step 2. Type the bank's website address.	3.01
Step 3. Verify that the bank's home page is displayed.	1.21
Total Execution Time	5.13

Method for Goal: Log into the System	Sub-Execution Time(s)
Step 1. Locate the "Online Banking" logon section on the bank's home page.	1.21
Step 2. Read your online ID that has been already filled in by the system (e.g. go4t****).	1.21
Step 3. Locate the "Sign In" button.	1.21
Step 4. Move cursor to "Sign In" button.	1.11
Step 5. Click on "Sign In" button with left mouse button	0.21
Step 6. Verify that it is a secure connection by checking if the bank's address bar contains the prefix "https" when the "Confirm that your SiteKey is correct" page is displayed.	1.21
Step 7. Verify that it is a secure connection by checking that the URL is correct and no errors were encountered on the same page.	1.21
Step 8. Think of the closed yellow padlock icon on the bottom right of the Windows task bar as indicating that it is a secure connection.	0.29
Step 7. Locate the SiteKey phrase field.	1.21
Step 8. Read the SiteKey phrase.	1.51
Step 9. Recognize the SiteKey phrase (e.g. "Whales are fascinating creatures"), retrieve it from LTM and retain it.	1.21
Step 10. Verify that the SiteKey phrase is correct according to the user's online account setup.	1.21
Step 11. Read the SiteKey image.	1.21
Step 12. Recognize your SiteKey image (e.g. "a whale image"), retrieve it from LTM and retain it.	1.21
Step 13. Verify that the SiteKey image (e.g. "a whale image") is correct according to the user's online account setup.	1.21
Step 14. Locate the "Passcode" field.	1.21

Step 15. Verify that the cursor has been already placed in the "Passcode" field on the page load.	1.21
Step 16. Recall the passcode (e.g. Light7ocean), retrieve it from LTM, and retain it.	1.21
Step 17. Type the passcode within the "Passcode" field.	0.51
Step 18. Verify that asterisks are displayed while entering the passcode within the "Passcode" field.	1.21
Step 19. Verify that passcode has been correctly typed in.	1.21
Step 20. Forget passcode.	1.21
Step 21. Locate the "Sign In" button.	1.21
Step 22. Click on "Sign In" button with left mouse button.	0.21
Step 23. Verify that the system has been granted access and the "Accounts" page is displayed.	1.21
Total Execution Time	26.83

Method for Goal: Make an Electronic Funds Transfer	*Sub-Execution Time(s)*
Step 1. Verify that you are in the "Accounts Overview" tab within "Accounts" page.	1.21
Step 2. Locate the "Transfers" tab and click on it.	1.21
Step 3. Verify that you are in the "Transfers" page.	1.21
Step 4. Verify that you are in the "Make Transfer" sub-tab.	1.21
Step 5. Locate the "From" field and select the account.	1.21
Step 6. Locate the "To" field and select the account.	1.21
Step 7. Locate the "Amount" field and type the amount of $15,000.	1.21
Step 8. Locate the "Continue" button and click on it.	1.21

Step 9. Read a warning message on the top of the next page which says that "You will have to provide an additional authentication credential in order to proceed with transfers over $10,000. Do you want to continue?".	1.21
Step 10. Verify the "OK" and "Cancel" buttons.	1.21
Step 11. Locate the "OK" button and click on it.	1.21
Step 12. Verify that you are in the "Additional Authentication Credentials" page.	1.21
Step 13. Locate the instructions section on this page.	1.21
Step 14. Read instructions regarding the four-digit number code that has to be sent to the iPhone mobile device using SMS. After 15 seconds, verify that an SMS text message has been sent by the bank to the iPhone mobile device. Then enter this same code in the "Code" field on this Web page.	10.01
Step 15. Wait for a sound alert for SMS notification to the iPhone after 15 seconds, indicating that the four-digit number code has been sent by the bank.	15.00
Step 16. Hear the auditory stimulus in the form of a sound alert for SMS notification with the four-digit number code that has been sent by the bank to the iPhone, retrieve it from the auditory LTM, and retain it.	1.21
Step 17. Auditory stimulus information (sound alert) dissipates from LTM and is no longer available.	1.01
Step 18. Open the "Messages" application on the iPhone.	2.01
Step 19. Verify that you have received a SMS text message with the four-digit number code sent by the bank.	1.21
Step 20. Read the four-digit number code directly from the subject field of the message row.	1.21
Step 21. Memorize the four-digit number code, and retain it in the STM.	6.00
Step 22. Locate the "Code" field on the "Additional Authentication Credentials" web page.	1.21

Step 23. Type the four-digit number code in the "Code" field.	2.16
Step 24. Verify that asterisks are displayed while entering the code within the "Code" field.	1.21
Step 25. Forget code.	1.21
Step 26. Verify the "Send" and "Cancel" buttons are displayed below the "Code" field on the page.	1.21
Step 27. Click on "Send" button with left mouse button.	0.21
Step 28. Verify that you have been directed to the confirmation page which states that you have successfully transferred the amount of $15,000 to the desired destination account.	1.21
Total Execution Time	61.81

The total execution time for **TASK_SCENARIO: T3 Transfer $15,000 to the Bank of America** is shown in the following:

TASK_SCENARIO: T3 Transfer $15,000 to the Bank of America	
Methods	*Sub-Execution Time(s)*
Method for goal: Go to the bank website	5.13
Method for goal: Log into the system	26.83
Method for goal: Make an electronic funds transfer	61.81
Total Execution Time	93.77

5.4.3.3.5 Estimate Total Learning Time

The total learning time for **TASK_SCENARIO: T3 Transfer $15,000 to the Bank of America** is shown in the following:

Total Learning Time = Pure Method Learning Time

+ LTM Learning Time

+ Training Procedure Execution Time

Total Learning Time = 833 secs + 102 secs + 0 secs = 935 secs

5.4.3.4 TASK: Access a File on a Personal Laptop

TASK_SCENARIO: T4	Access a file on a personal laptop
TASK CATEGORY	Wireless network and electronic access control-based task.
USER AUTHENTICATION METHOD	Fingerprint recognition (biometrics)

5.4.3.4.1 Generate Task Description

This task uses a portable USB device that allows, for instance, remote employees to swipe their finger to access corporate network resources. It recognizes a fingerprint, providing a secure way of accessing a protected resource. Users have to install the USB fingerprint suite software on the desktop or laptop computer; plug the USB fingerprint into the USB port, and, lastly, swipe their finger on the fingerprint reader to log into Windows, for example, or to access password-protected websites. A pre-condition is that users have to set up a one-time registration for all their accounts to authenticate with the USB, but once that is set up, users can make use of the USB fingerprint.

5.4.3.4.2 Describe a List of High-Level User Goals

The user's topmost goal is: Access a file on a personal laptop. The set of the user's high-level goals considered are:

- Log into the system.
- Go to the file directory.
- Open the file.
- Return with goal accomplished.

5.4.3.4.3 Define Operators and Write Methods for Accomplishing User Goals

- **Method for goal: Log into the system.**
  Step 1. Read fingerprint logon Welcome screen containing finger image and "Password" field on the laptop computer.
  Step 2. Refer to the USB-based biometric fingerprint reader.
  Step 3. Locate the fingerprint sensor on the USB fingerprint reader.
  Step 4. Move finger to the USB fingerprint reader.
  Step 5. Place last knuckle joint over the center of the fingerprint sensor.
  Step 6. Swipe the finger without lifting it over the fingerprint sensor and read the computer screen.

Step 7. Verify that you have been granted access to the system.

Step 8. Return with goal accomplished.

■ **Method for goal: Go to the file directory.**

Step 1. Open the file manager application (e.g. Windows or MAC Explorer) on the laptop computer.

Step 2. Verify that the Explorer application is opened.

Step 3. Locate the C: directory on the Explorer.

Step 4. Move cursor to the C: directory and click on it.

Step 5. Verify that the Explorer application is on the C: directory.

Step 6. Return with goal accomplished.

■ **Method for goal: Open the file.**

Step 1. Locate the "read.txt" file on C: directory.

Step 2. Move cursor to the "read.txt" file icon.

Step 3. Double-click on "read.txt" file icon.

Step 4. Verify that the "read.txt" file is opened.

Step 5. Access the "read.txt" file.

Step 6. Return with goal accomplished.

5.4.3.4.4 Estimate Total Execution Time

The time measurement related to the *execution time* for **TASK_SCENARIO: T4 Access a file on a personal laptop** is listed in seconds in the following.

Method for Goal: Log into the System	Sub-Execution Time(s)
Step 1. Read fingerprint logon Welcome screen containing finger image and "Password" field on the laptop computer.	2.11
Step 2. Refer to the USB-based biometric fingerprint reader.	1.10
Step 3. Locate the fingerprint sensor on the USB fingerprint reader.	1.21
Step 4. Move finger to the USB fingerprint reader.	1.11
Step 5. Position last knuckle joint over the center of the fingerprint sensor.	1.11
Step 6. Swipe the finger without lifting it over the fingerprint sensor.	1.31
Step 7. Verify that you have been granted access to the system.	1.21
Total Execution Time	9.16

Method for Goal: Go to the File Directory	Sub-Execution Time(s)
Step 1. Open the file manager application (e.g. Windows or MAC Explorer) on the laptop computer	0.41
Step 2. Verify that the Explorer application is opened	1.21
Step 3. Locate the C: directory on the Explorer	1.21
Step 4. Move cursor to the C: directory and click on it	1.11
Step 5. Verify that the Explorer application is on the C: directory	1.21
Total Execution Time	5.15

Method for Goal: Open the File	Sub-Execution Time(s)
Step 1. Locate the "read.txt" file on C: directory.	2.21
Step 2. Move cursor to the "read.txt" file icon.	1.11
Step 3. Double-click on "read.txt" file icon.	0.41
Step 4. Verify that the "read.txt" file is opened.	1.21
Step 5. Access the "read.txt" file.	1.21
Total Execution Time	6.15

The total execution time for **TASK_SCENARIO: T4 Access a file on a personal laptop** is shown in the following:

TASK_SCENARIO: T4 Access a File on a Personal Laptop	
Methods	Sub-Execution Time(s)
Method for goal: Log into the system	9.16
Method for goal: Go to the file directory	5.15
Method for goal: Open the file	6.15
Total Execution Time	20.46

Estimate Total Learning Time:

The total learning time for **TASK_SCENARIO: T4 Access a file on a personal laptop** is shown in the following:

Total Learning Time = Pure Method Learning Time

+LTM Learning Time

+Training Procedure Execution Time

Total Learning Time = 289 secs + 48 secs + 0 secs = 337 secs

5.4.4 Time-Level Analysis of NGOMSL

This section presents a time analysis of the data gathered for each set of four task scenarios. As described previously, the design information obtained from NGOMSL has been the operator sequences, execution times, and procedure learning times.

As shown in Table 5.1, the user took 28.85 seconds, which is the total execution time, to check business email (T1) using the password/PIN authentication method and so forth for the tasks T2, T3, and T4. It is worth mentioning that what is important is not to measure the TET the user has spent in each task as a whole but rather just the TET to "Log into the system". It is irrelevant to only measure how long it would take, for instance, to check business email (i.e. 28.85 seconds), update a specification (i.e. 45.31 seconds), make an electronic funds transfer (i.e. 93.77 seconds), or access a file (i.e. 20.46 seconds).

These tasks can significantly vary in terms of time, application type, and the context in which they have been performed. However, the TET does vary depending on the type of the authentication method used. In fact, it takes more time if the user employs an OOBA method rather than another method because the amount

Table 5.1 Total Execution Time by Task Scenario

Task Scenario	Description	Authentication Method	Total Execution Time(s)
T1	Check business email	Password/PIN	28.85
T2	Update the SecurID token UI spec	OTP	45.31
T3	Transfer 15,000 to the Bank of America	OOBA	93.77
T4	Access a file on a personal laptop	Fingerprint	20.46

of user interaction when authenticating to a system with OOBA is more demanding than the other authentication methods.

The results show the total execution time for the set of four authentication benchmark methods, which is the profile for the Method for goal: Log into the system in Table 5.2. The profile includes the total time in seconds spent using this method and the percentage of the time spent on it. What the authors are more concerned with is an investigation of the authentication elements of the tasks scenarios. These elements are in fact the time related to the Method for goal: Log into the system.

As previously mentioned, the main factors influencing the amount of time a user spends authenticating to a system are the number of different artifacts to interact with and the authentication method type. For instance, in the OOBA method, Alice needs to interact with the bank's website and then a mobile device in order to accomplish the Method for goal: Log into the system. Also, the authentication method type such as password/PIN takes more time to be performed (23.25 seconds) when compared to fingerprint recognition (9.16 seconds). The former is a knowledge-based authentication (KBA), which requires users to prove the knowledge of a single secret, memorize items, and recall them when accessing a specific system. On the other hand, the latter is biometrics, which recognizes users physically through their fingers; no cognitive process is directly involved.

Using OTP takes a little more time than OOBA because users need to both interact with different artifacts and make use of KBA, which directly involves cognitive processes. With OTP, users are required to refer to a hardware authentication token, then type the code displayed in the token on their application (e.g. VPN application). In addition, users need to remember the PIN (i.e. four-digit) but not the password (i.e. strong password like Rtyr78nM!), which facilitates memory retrieval, although this authentication method is the one that takes more time.

As expected, the fingerprint authentication method (biometrics) takes the least amount of time out of all the methods. No cognitive process is directly involved (e.g. not KBA), and there is minimal interaction with artifacts when using a USB drive. It is important to note that it is the acquisition of the user credentials (e.g.

Table 5.2   Execution Time by Task Scenario, Authentication Method Type, and Method for Goal: Log into the System

Task Scenario	Method for Goal	Authentication Method	% of Total	Total Execution Time(s)
T1	Log into the system	Password/PIN	83.93	23.25
T2		OTP	12.75	28.13
T3		OOBA	25.16	26.83
T4		Fingerprint	1.88	9.16

the biometric reader captures fingerprint samples from users) that takes less time when compared with the other authentication methods presented in Table 5.2. The exact authentication processing time can vary considerably depending on the infrastructure, the equipment, and also on different versions of (the same) authentication methods.

5.5 A Concluding Remark

In this chapter, we introduced the key stages of the usable security protocol methodology while further detailing the first three stages: Define, Identify, and Develop. The three first stages are:

- Define the mission and conceptual design objective, including, a clear definition of usable security problems, task scenario, usability scenario and security ones and well as the users and their working contexts.
- Identify the most representative user authentication methods used by users and in which the usable security problems occurred.
- Develop the Natural GOMS Language. This consists of describing precisely the tasks for each class of users using as input the task, usability, and security scenarios. The description of GOMS include also the measures that can be used to assess the degree of conflict between usability and security.

The next chapter details the stages: Assess and Generate.

Chapter 6

The Usable Security Protocol Methodology: Assess and Generate

In Chapter 5, we presented the first three steps of the methodology, namely define, identify, and develop; here we present the next steps of the methodology: assess and generate.

6.1 Develop the Authentication Risk-Assessment Matrix

If we don't identify where the most critical vulnerabilities are – when related either to security or usability – then how can we secure our system's information infrastructure? The Authentication Risk-Assessment Matrix must be developed prior to the development of the usable security inspection method itself. It is, in fact, a crucial step in acknowledging and understanding the main threats and security vulnerabilities related especially to online user authentication (user-to-machine). It determines which "Security Review" should be considered within each "Usability Criterion" in the usable security symmetry inspection method. The primary attacks considered by the authors are the following: Eavesdropping, Man-In-The-Middle (MITM), Replay, Session hijacking, and Verifier impersonation attacks.

The Open Web Application Security Project (OWASP) is an open-source application security project. The OWASP community consists of corporations, educational organizations, and a variety of security experts worldwide who share their knowledge of vulnerabilities, threats, attacks, and countermeasures.

This community works to generate freely available articles, methodologies, documentation, tools, and technologies. OWASP is not affiliated with any technology organization or company, although it supports the knowledgeable use of security technology. OWASP has avoided affiliation as it believes freedom from organizational pressures may make it easier for it to delivoffer impartial, practical, cost-effective information about application security. As a matter of fact, the U.S. Federal Trade Commission strongly recommends that organizations use the OWASP Top Ten and ensure that their partners do the same. In addition, the U.S. Defence Information Systems Agency has listed the OWASP Top Ten as key best practices that should be used as part of the United States of America (USA) Department of Defense (DoD) Information Technology Security Certification and Accreditation Process (DITSCAP). For these reasons, the authors adopted OWASP as the primary source for identifying the most critical security vulnerabilities.

The OWASP Top Ten provides a powerful awareness document and minimum standard for web application security (OWASP, 2009). It represents a broad consensus on the most critical web application security flaws, including authentication. For more information regarding the methodology used by OWASP to select the security vulnerabilities, go to: https://www.owasp.org/index.php/Top_10-2017_Methodology_and_Data.

It is worth noting that five out of the top ten security vulnerabilities are directly or indirectly related to authentication. The top ten most critical web application security vulnerabilities are described as follows (NOTE: the symbol "✪" located next to the security vulnerability name means that it is related to authentication):

- *Invalidated input*: Information from web requests is not validated before being used by a web application. Attackers can use these flaws to attack back end components through a web application.
- *Broken access control* ✪: Restrictions on what authenticated users are allowed to do are not properly enforced. Attackers can exploit these flaws to access other users' accounts, view sensitive files, or use unauthorized functions.
- *Broken authentication and session management* ✪: Account credentials and session tokens are not properly protected. Attackers that can compromise passwords, keys, session cookies, or other tokens can defeat authentication restrictions and assume other users' identities.
- *Cross Site Scripting (XSS) flaws* ✪: The web application can be used as a mechanism to transport an attack to an end user's browser. A successful attack can disclose the end user's session token, attack the local machine, or spoof content to fool the user.
- *Buffer overflows*: Web application components in some languages that do not properly validate input can be crashed and, in some cases, used to take control of a process. These components can include Common Gateway Interface (CGI), libraries, drivers, and web application server components.
- *Injection flaws* ✪: Web applications pass parameters when they access external systems or the local operating system. If an attacker can embed malicious

commands in these parameters, the external system may execute those commands on behalf of the web application.

■ *Improper error handling*: Error conditions that occur during normal operations are not handled properly. If an attacker can cause errors to occur that the web application does not handle, they can gain detailed system information, deny service, cause security mechanisms to fail, or crash the server.

■ *Insecure storage*: Web applications frequently use cryptographic functions to protect information and credentials. These functions and the code to integrate them have proven difficult to code properly, frequently resulting in weak protection.

■ *Denial-of-service (DOS)*: Attackers can consume web application resources to the point where other legitimate users can no longer access or use the application. Attackers can also lock users out of their accounts or even cause the entire application to fail.

■ *Insecure configuration management* ✪: Having a strong server configuration standard is critical to a secure web application. These servers have many configuration options that affect security and are not secure out of the box.

The following paragraphs present a sample of the Authentication Risk-Assessment Matrix (Table 6.1), which describes the user authentication assets, threats, and vulnerabilities along with their corresponding descriptions and mitigation strategies, types of rating scales for Threat, Vulnerability, CIA (Confidentiality, Integrity, and Authorization model), Probability, Asset Value and Asset Exposure Classifications, Total Impact and Total Risk Ratings, and finally Risk Reduction Strategy. For the full description of all Authentication Assets/Targets, please see "Appendix 1: Authentication Risk-Assessment Matrix."

Matrix Legend

✪	=	A specific authentication/OWASP top ten related security vulnerability.
T	=	Threat.
V	=	Vulnerability.
CP/E	=	Compromise/Exploit.
RA	=	Risk Assessment.
OP	=	Overall Probability Rating is the sum of the Threat and the Vulnerability Rating (OP=T+V).
TI	=	Total Impact Rating is the sum of the Asset Value Classification and the Asset Value Exposure (TI=AVC+AVE).
TR	=	Total Risk Rating is the product of the Overall Probability and the Total Impact (TR=OP×TI).

Table 6.1 Authentication Risk-Assessment Matrix

Authentication Asset/Target	Threat (T) Description	Vulnerability (V) Description	Overall PC/E				Overall Exposure/Impact			RA	Risk Reduction Strategy
			CIA	Threat Rating	V Rating	OP=T+V	Asset Value Classification	Asset Value Exposure	TI=AVC+AEC	TR=OP×TI	
1. PASSWORD Personally identifiable medical Information stored on *Structured Query Language* (SQL) Server	Account credentials of data entry clerk stolen	Overly complex password requirements cause users to write down passwords and leave them in obvious places.	CIA	3 Medium	3 Medium	6	4 Substantial	4 Serious	8	48	Mitigate by reducing password complexity requirements, enforcing policy to not leave passwords in obvious places, and providing user training in password use. Compromise of SQL data could result in large fines as a result of HIPAA[a] violations. Also, loss of public confidence could result in long-term loss of business.
2. PASSWORD Personally identifiable medical Information stored on SQL Server	Help desk resets password for an account used by data entry clerk based on request from unauthorized individual (Social Engineering attack[b]).	Lack of policies and procedures in place to verify identity of individual requesting password reset.	CIA	2 Low	2 Medium	4	4 Substantial	4 Serious	8	24	Mitigate by implementing tighter procedures to verify identity–and by delegating password change permissions to small business units where requests for password changes come from individuals known to administrator. Compromise of SQL data could result in large fines as a result of HIPAA violations. Also, loss of public confidence could result in long-term loss of business.

a The Health Insurance Portability and Accountability Act of 1996 (HIPAA) Privacy Rule http://www.hhs.gov/ocr/privacy/hipaa/understanding/index.html

b A social engineering attack is one in which the intended victim is somehow tricked into doing what the attacker requests.

■ Definitions

A *threat* is a circumstance, event, or person with the potential to cause harm to a system in the form of destruction, disclosure, data modification, and/or Denial-of-Service (DoS).

Vulnerability is a hole or a weakness in the application, which can be a design flaw or an implementation bug that allows an attacker to cause harm to the stakeholders of an application. Stakeholders include the application owner, application users, and other entities that rely on the application.

Risk Assessment is a computation of risk. Risk is a threat that exploits some vulnerability that could cause harm to an asset. The risk algorithm computes the risk as a function of the assets, threats, and vulnerabilities. One instance of a risk within a system is represented by the formula RISK = ASSET × THREAT × VULNERABILITY. Total risk for a network equates to the sum of all the risk instances.

■ Ratings:

The ratings described by Shadow (2008) in the succeeding paragraphs have been applied in the Authentication Risk-Assessment Matrix.

■ Threat, Vulnerability, and Overall Probability Scales:

Threat Rating	
1.	Very low or negligible probability of threat. Little or no motivation to launch attack. Almost no probability of threat for non-human threat agent.
2.	Low probability of threat.
3.	Medium probability of threat.
4.	High probability of threat.
5.	Extremely high, almost certain probability of threat.

Vulnerability Rating	
1.	Very low or negligible. Vulnerability requires extensive effort/knowledge/resources to exploit; exploit of vulnerability does not lead to exposure of additional vulnerabilities in other services/systems/processes.
2.	Low. Vulnerability requires significant effort/knowledge/resources to exploit. Low probability that exploit will create exposure to additional vulnerabilities in and threats to other services/systems/processes.

3.	Medium. Vulnerability requires a moderate amount of effort/knowledge/resources to exploit. Moderate probability that exploit will create exposure to additional vulnerabilities in and threats to other services/systems/processes.
4.	High. Vulnerability requires some resources to exploit. High probability that exploit will create exposure of additional vulnerabilities in and threats to other systems/services/processes.
5.	Very High. Vulnerability requires little knowledge, effort, or skills to exploit. Very high probability that exploit will create exposure of additional vulnerabilities in and threats to other systems/services/processes.

Overall Probability Matrix (Threat/Vulnerability)						
Threat Rating	Vulnerability:					
	0	1	2	3	4	5
Very Low	1	2	3	4	5	6
Low	2	3	4	5	6	7
Medium	3	4	5	6	7	8
High	4	5	6	7	8	9
Very High	5	6	7	8	9	10

■ Exposure/Impact Rating Scales

Asset Value Classification	
1.	Negligible asset value. Negligible or no impact on the business if confidentiality or integrity of asset is compromised. Compromise of availability results in negligible or no increase of support costs or loss of productivity.
2.	Low asset value. Low impact on business that cannot be measured if confidentiality or integrity of asset is compromised. Compromise of availability results in distractions that are easily absorbed by the internal business process – possible slight increase in support costs.

3.	Medium asset value. Medium impact on business (internal processes, etc.) if confidentiality or integrity is compromised, resulting in revenue loss and increase in support costs. Compromise of availability results in work delays with a noticeable increase in support costs and loss of productivity.
4.	Substantial asset value. Serious impact on the business if confidentiality or integrity of asset is compromised, resulting in loss of profitability or success. Compromise of availability results in work interruptions, causing a quantifiable increase in support costs or delay in business commitments (e.g. clients and customers are unable to connect to websites, unable to make commitments for contract deliverables on time, etc.).
5.	High asset value. Severe or catastrophic impact on the business if the confidentiality of assets is compromised, resulting in high losses to business profitability or success. Compromise of availability results in significant work stoppages, causing a substantial increase in support costs or cancellation of business commitments.

Exposure Classification	
1.	Negligible or no loss of asset confidentiality, integrity, or availability. Effects of compromise to asset severely contained with no subsequent threat of compromise to other assets.
2.	Low loss of asset confidentiality, integrity, or availability. Effects of compromise to assets tightly contained with negligible or low subsequent threat to other assets.
3.	Moderate or limited loss of asset confidentiality, integrity, or availability. Effects of compromise to assets can involve more than one system or service and cause an increased threat to other assets. Compromise or exploit may be externally visible.
4.	Serious loss of asset confidentiality, integrity, or availability. Effects of compromise are likely to have negative effects on other assets and cause a noticeable increase in threats to other assets. Compromise or exploit may be externally visible.
5.	Severe or complete loss of asset confidentiality, integrity, or availability. Results in a significant increase in threats to other assets. High probability compromise or exploit may be externally visible.

■ Total Impact Matrix

Asset Value Classification	Exposure Factor: 1=Negligible, 2=Low, 3=Medium, 4=Serious, 5=Severe					
	0	1	2	3	4	5
1=Very low or negligible	1	2	3	4	5	6
2=Low	2	3	4	5	6	7
3=Medium	3	4	5	6	7	8
4=Substantial	4	5	6	7	8	9
5=High	5	6	7	8	9	10

■ Overall Risk Matrix (Overall Probability × Total Impact)

HIGHEST IMPACT OF THE RISK EVENT

Probability x Highest Impact = Risk Severity Index

PROBABILITY									
1	2	3	4	5	6	7	8	9	10
2	4	6	8	10	12	14	16	18	20
3	6	9	12	15	18	21	24	27	30
4	8	12	16	20	24	28	32	36	40
5	10	15	20	25	30	35	40	45	50
6	12	18	24	30	36	42	48	54	60
7	14	21	28	35	42	49	56	63	70
8	16	24	32	40	48	56	64	72	80
9	18	27	36	45	54	63	72	81	90
10	20	30	40	50	60	70	80	90	100

Risk Summary Ranges	
Low	1–19
Medium	20–40
High	41–100

Additional explanation of some of the security vulnerabilities mentioned in the Authentication Risk-Assessment Matrix is provided in the next section.

6.1.1 Common Security Exploits Method

■ **Cross Site Scripting (XSS)***

XSS is generally believed to be one of the most common application layer hacking techniques. Generally speaking, XSS refers to a hacking technique that leverages vulnerabilities in the code of a web application to allow an attacker to send malicious content from an end-user and collect some type of data from the victim. Today, websites rely heavily on complex web applications to deliver different output or content to a wide variety of users according to set preferences and specific needs. This arms organizations with the ability to provide better value to their customers and prospects. However, dynamic websites suffer from serious vulnerabilities, rendering organizations helpless and prone to cross site scripting attacks on their data. A web page contains both text and HTML markups that are generated by the server and interpreted by the client browser. Websites that generate only static pages are able to have full control over how the browser interprets these pages. Websites that generate dynamic pages do not have complete control over how their outputs are interpreted by the client. The heart of the issue is that if mistrusted content can be introduced into a dynamic page, neither the website nor the client has enough information to recognize that this has happened and take protective actions. XSS allows an attacker to embed malicious JavaScript, VBScript, ActiveX, HTML, or Flash into a vulnerable dynamic page to fool the user, executing the script on his machine in order to gather data. The use of XSS might compromise private information, manipulate or steal cookies, create requests that can be mistaken for those of a valid user, or execute malicious codes on the end-user systems. The data is usually formatted as a hyperlink containing malicious content and which is distributed over any possible means on the Internet.

■ **Buffer Overflows**

Buffer overflow exploits constitute the largest single threat to enterprises today. These exploits have the most power, are the easiest to use, and are all too common. No advanced technical knowledge is necessary to run pre-written buffer overflow exploit codes. Buffer overflow exploits are very powerful, and in many cases, the malicious code that executes as a consequence of a buffer overflow will run with administrator level privileges, and thus can do anything it wants to the server.

---

* Cross Site Scripting Attack. Web Site Security. Acunetix, Inc. September 1st, 2018 http://www.acunetix.com/websitesecurity/cross-site-scripting.htm.

■ **Broken Authentication**

For example, when a user provides his login name and password to authenticate and prove his identity, the application assigns the user specific privileges to the system based on the identity established by the supplied credentials. Hijacking control of a connection is taken by the attacker after the user authentication has been established. This kind of attack is not a technological security hole in the Operating System or server software, but rather, it depends on how securely stored and complex the passwords are and on how easy it is for the attacker to reach the server (network security).

■ **Password Guessing**

Password guessing can be one of the most efficient techniques to defeat web authentication. This technique can be carried out either manually or via automated procedures. Table 6.2 shows some common usernames and passwords used by attackers in authentication guessing attacks:

■ **Brute-Force Attack**

A Brute-Force Attack is the most widely known password cracking method. If password guessing renders no results, the next step for an attacker is to try other password combinations using special custom tools, such as WebCracker (https://www.symantec.com/security-center/writeup/2004-121612-5341-99), which is freely available on the Internet. This custom tool attempts to authenticate into the system, making use of predefined lists of usernames and passwords, dictionary attacks, and brute-force attacks. A dictionary attack uses pre-computed wordlists like dictionaries to try to authenticate on the web applications by trying thousands of combinations of these dictionary words as usernames and passwords.

Table 6.2 Username and Password Guessing (Scambray et al., 2006)

Username Guessing	Password Guessing
(NULL)	(NULL)
root, administrator, admin	(NULL), root, administrator, admin, password (company name)
operator, webmaster, backup	(NULL), operator, webmaster, backup
guest, demo, test, trial	(NULL), guest, demo, test, trial
member, private	(NULL), member, private
(company name)	(NULL), (company name), password
(company name)	(NULL), (known name)

■ LanMan (LM) Hash Algorithm

LM Hash is a compromised password hashing function that was the primary hash that Microsoft LAN Manager and Microsoft Windows versions prior to Windows NT used to store user passwords. One setback with this encryption scheme is that all characters are converted to uppercase prior to encryption. This, in fact, removes 26 characters from the set of choices from which a user may possibly select a password, making a dictionary attack, or even a brute-force attack, considerably less work for a cracker. Another weakness of the LM Hash scheme is an even greater one, however, because of the method used to prepare the password for encryption. The number of characters in an LM password is exactly 14, no matter how many characters a user chooses. Perhaps a 14-character password seems like a good one, but this is not the case. Each user password of fewer than 14 characters is padded with null characters (ASCII zero) to extend its length. The result is then split into two 7 character parts, each of which is encrypted separately. Along with a predictable parity value, the results are hashed, concatenated, and stored.

■ Password History and Password Aging

Password expiration is not efficient unless users choose different passwords from those previously used. *Password history* is the retention of one or more prior passwords or password hashes for comparison against new passwords or password hashes. A new password is checked to make sure that it has not been used during the specified history. The period is typically defined as either a certain number of prior passwords or a period of time. *Password age*, in turn, is an attribute directly related to password history. The *minimum password age* is the amount of time that must pass between password changes. As Scarfone and Souppaya (2009) point out, to diminish the effort necessary in remembering passwords, a significant number of users will cycle through passwords after expiration until they have exceeded the password history retention buffer and then change their password back to the original one. Although enforcing a minimum password age does not prevent this, at least it is a restriction.

There are some password history mechanisms that are also capable of identifying passwords that are not satisfactorily different from previous passwords. When forced to choose a new password, the majority of users have a tendency to employ variations of old passwords (e.g. changing "secret05" to "secret06"). This makes it easy for an attacker who knows the old password to guess or crack the new one rapidly. Some existing password history mechanisms can be configured to refuse new passwords that have a certain number of characters in common with previous passwords. Without such a mechanism, it is usually trouble-free for users to append counters to their passwords (e.g. "secret05"). This makes password expiration mostly unproductive, and may, in fact, cause users to select weaker passwords than they would have without password expiration.

Password history usually only works on a single authentication mechanism and cannot check the history from multiple mechanisms. This enables users to employ the identical password (and prior passwords) on several systems at once. Users frequently do this because it decreases the number of passwords that they have to remember, but increases the risk to the enterprise by entitling an attacker who compromises one password to reuse it to gain access to additional resources. Additionally, administrators will sometimes reuse passwords between a local user account on a personal workstation and an account that has domain or centralized administrative privileges. This can pose a major risk to the enterprise because the security of centralized password management is generally higher than on individual workstations. An attacker who compromises the workstation and is able to crack the domain administrator password will have significant access to enterprise resources.

There is generally no easy way to detect password reuse across systems, particularly when both internal and external systems are involved. To attempt to reduce the likelihood of password reuse, organizations can have their password management policies prohibit the use of the same or closely related passwords on the organizational IT system and external systems. The password management policy can also explicitly forbid the reuse of centralized (e.g. domain) administrative level credentials with user or local (e.g. local administrator or root) accounts. Proper user training that stresses the importance of proper password management and protection and explains the risks of password reuse should also be implemented. However, without an enforcement mechanism, it is unlikely that policies against reuse will be significantly effective in reducing reuse, given the number of passwords that users typically need to remember.

■ Shoulder Surfing

Shoulder surfing is using direct observation techniques, such as looking over someone's shoulder, to get information. Shoulder surfing is an efficient way to get information in packed places because it's quite simple to stand next to someone and watch as they fill out a form, enter a PIN number at an Automated Teller Machine (ATM), or use a calling card at a public pay phone. Shoulder surfing can also be done long distance with the assistance of binoculars or other vision-enhancing devices. To prevent shoulder surfing, experts advise that you protect paperwork or your keypad from view by using your body or hand.

■ Phishing

Phishing is the criminally fraudulent process of attempting to obtain sensitive information such as usernames, passwords, and credit card details by masquerading as a trustworthy entity in an electronic communication. Phishing is typically carried out by email or instant messaging. A typical example is an email that directs users to visit a website where they are asked to update personal information, such as passwords and credit card, social security, and bank account numbers that the legitimate organization already has. The website, however, is forged and set up only to steal the user's information. Phishing is one of the social engineering techniques employed to trick users and exploits the poor usability of existing web security technologies.

■ Hard-Coded Password

Using a hard-coded password greatly increases the possibility of password guessing. The consequences in authentication are the following: If hard-coded passwords are used, it is almost certain that attackers will gain access through the account in question. They continue to be used to this day, sometimes in high-profile software, despite the significant risk they pose. An example of a hard-coded password in the Java programming language is shown in Figure 6.1. The 8-digit characters Boat6sea! is the hard-coded password.

■ Trojan Horse

A Trojan horse is a rogue program that takes the identity of a trusted application to collect information or avoid detection. In a typical Trojan horse attack, the user is presented with a logon screen that appears to be genuine. The user enters their user name and password, and are either logged on, or presented with an error message that she has to type their logon credentials again. Often, the rogue logon application exits after the first request passed the user on to the real logon. Users are easily fooled into thinking that they probably typed the wrong password and must re-enter the information again, never suspecting that their logon credentials are compromised.

■ Man-In-The-Middle Attack

In a Man-In-The-Middle (MITM) attack, a malicious party intercepts a legitimate communication between two friendly parties. The malicious host then controls the flow of communication and can eliminate or alter the information sent by one of the original participants without the knowledge of either the original sender

```
intVerifyAdmin(String password) {
        if (passwd.Equals("Boat6sea!")) {
        return(0)
        }
        return(1);
}
```

Figure 6.1   Hard-coded password (Boat6sea!): JAVA code snippet.

or the recipient. In this way, an attacker can fool a victim into disclosing confidential information by "spoofing" the identity of the original sender, who is presumably trusted by the recipient. Here is an example of a MITM attack against a Web-based financial system: A bank demands authentication from the user (i.e. a password, a one-time code from a token, etc.). The attacker sitting in the middle receives the request from the bank and passes it on to the user. The user responds to the attacker, who passes that response to the bank. Now the bank assumes it is talking to the legitimate user, and the attacker is free to send transactions directly to the bank. This kind of attack completely bypasses any two-factor authentication mechanisms and is becoming a more popular identity-theft tactic.

■ Broken Authentication

User authentication on the Web usually involves the use of a user ID and password. Stronger methods of authentication are commercially available, such as software- and hardware-based cryptographic tokens or biometrics, but such mechanisms are cost prohibitive for most web applications. A wide array of account and session management flaws can result in the compromise of user or system administration accounts. Development teams frequently underestimate the complexity of designing an authentication and session management scheme that adequately protects credentials in all aspects of the site. Web applications must establish sessions to keep track of the stream of requests from each user. HTTP* does not provide this capability, so web applications must create it themselves. Frequently, the web application environment provides a session capability, but many developers prefer to create their own session tokens. In either case, if the session tokens are not properly protected, an attacker can hijack an active session and assume the identity of a user. Creating a scheme to create strong session tokens and protect them throughout their lifecycle has proven obscure for many developers. Unless all authentication credentials and session identifiers are protected with SSL† at all times and protected against disclosure from other flaws, such as XSS, an attacker can hijack a user's session and assume their identity.

■ SQL Injection

SQL Injection attacks are very common, and this is due to two factors: the significant prevalence of SQL Injection vulnerabilities and the attractiveness of the

---

* Hypertext Transfer Protocol (HTTP) is an application-level protocol for distributed, collaborative, hypermedia information systems. Its use for retrieving inter-linked resources led to the establishment of the World Wide Web.
† Secure Socket Layer (SSL) is a protocol for transmitting private documents via the Internet. SSL uses a cryptographic system that uses two keys to encrypt data – a public key known to everyone and a private or secret key known only to the recipient of the message. By convention, URLs that require an SSL connection start with "https://" instead of "http://".

target (e.g. the database usually contains all the appealing and critical data for your application). There are many successful SQL Injection attacks that occur because it is tremendously simple to introduce SQL Injection vulnerabilities in the code. Basically, SQL Injection flaws are introduced when software developers create dynamic database queries that include user supplied input. To avoid SQL Injection flaws is simple. Developers need to either stop writing dynamic queries and/or prevent user supplied input, which contains malicious SQL, from affecting the logic of the executed query.

■ **Privilege Escalation**

Privilege escalation occurs when a user gains access to more resources or functionality than they are normally allowed, and such elevation/changes should have been prevented by the application. This is typically caused by a flaw in the application. The result is that the application performs actions with more privileges than those intended by the developer or system administrator. The degree of escalation depends on which privileges the attacker is authorized to possess and which privileges can be obtained in a successful exploit. For example, a programming error that allows a user to gain extra privilege after successful authentication limits the degree of escalation because the user is already authorized to hold some privilege. Usually, we refer to *vertical escalation* when it is possible to access resources granted to more privileged accounts (e.g. an Internet banking account that acquires administrative privileges as an admin), and to *horizontal escalation* when it is possible to access resources granted to a similarly configured account (e.g. a user accesses information related to a different user on an online banking website).

■ **Real-Time Man-In-The-Middle Attacks: Session-Hijacking Trojans**

As mentioned in the item "15. OUT OF BOUND AUTHENTICATION (OOBA) Trojan Horse (TH)" in the Authentication Risk-Assessment Matrix (Appendix 1), this attack installs a type of proxy on the user's computer that interacts with the financial institution's genuine site on the user's behalf (i.e., in this type of MiTM attack, an attacker hijacks a session between a trusted client and a network server). As the Trojan interacts with the financial institution's site through the user's computer, it allows the fraudster to imitate the user's profile. Some categories of Trojans typically wait until the user logs onto the genuine site and perform a concurrent web session automatically. Thus the Trojan will appear to be transacting from the same IP and device as the user. This type of Trojan circumvents many existing security methods that rely on the compromised computer to communicate. In other words, it uses compromised devices to prompt users to supply challenge questions, one-time passwords, and other types of information that can be used to perpetrate fraud.

6.2 Generate the Usable Security Principles

This step mainly identifies and explains the main cognitive areas of focus relating to user authentication (such as perception, attention, memory, and other relevant areas). It also specifies a Cognitive Model of User Authentication (CMUA) in order to understand how and what cognitive processes (i.e., cognitive processes workflow) are involved for each of the main categories of user authentication methods. It serves as the basis for the development of the usable security inspection method.

Cognitive analysis for human–computer interaction (HCI) lends itself to two related interpretations: 1) the analysis of cognition-intensive interactions with computers, such as learning, problem-solving, or reading and 2) the analysis of cognitive content, structures, and processes involved in any interaction with a computer. This book addresses both interpretations by providing methods for analyzing cognition with a focus on interactions specifically related to user authentication involving cognitive processes such as Perception, Memory (Long-Term Memory (LTM), Short-Term Memory (STM), Visual Recognition Memory), Information Retrieval (Recall and Recognition), and Mental Models. In addition, the analysis of users' cognition should not be restricted to the early design phase (as task analysis typically is), but should be an important activity throughout the entire design process. However, many designers have little training in the methods used to measure cognition.

Security systems must be viewed as socio-technical systems that depend on the social context in which they are embedded to function correctly. Security systems will only be able to provide the intended protection when people actually understand and are able to use them correctly. There are very real differences between the degree to which systems can be considered theoretically secure (assuming they are correctly operated) and actually secure (acknowledging that often they will be operated incorrectly) (Jøsang et al., 2007b). In many cases, there is a trade-off between usability and theoretical security. It can be meaningful to reduce the level of theoretical security to improve the overall level of actual security. For example, the strongest passwords, from a theoretical perspective, are randomly generated. However, since it is very difficult to remember such passwords, people will write them down, and thereby undermine the system's security. Thus, it may be meaningful to allow people to choose passwords that are easier to remember. Although this reduces the theoretical strength of the passwords, it increases the security of the system as a whole.

It is important to step back and remind ourselves of the concept of cognition. The term cognition (Latin *cognoscere*, "to know" or "to recognize") is a concept used in different ways by different disciplines, but is generally accepted to mean the process of thought. It refers to a faculty for the processing of information, applying knowledge, and changing preferences. Within psychology or philosophy, the concept of cognition is closely related to abstract concepts such as the mind, reasoning, perception, intelligence, learning, and many others that describe capabilities of the mind and expected properties of an artificial "mind". Cognition is considered an abstract property of advanced living organisms and is studied as a direct property of a brain (or of an abstract mind) on the factual and symbolic levels. For example, in psychology and cognitive science, it refers to an information processing view of an individual's psychological functions.

Cognitive Informatics (CI) is an emerging discipline that studies the natural intelligence and internal information processing mechanisms of the brain, as well as the processes involved in perception and cognition. CI provides a coherent set of fundamental theories, in conjunction with contemporary mathematics, which form the foundation for most information and for knowledge-based science and engineering disciplines such as computer science, cognitive science, neuropsychology, systems science, cybernetics, software engineering, and knowledge engineering.

In psychology and in artificial intelligence, cognition is used to refer to the mental functions, mental processes (thoughts), and states of intelligent entities (humans, human organizations, highly autonomous machines). In particular, the field focuses on the study of specific mental processes such as comprehension, inference, decision making, planning, and learning.

The advent of (and constantly evolving) Information Technology (IT) has placed heavy cognitive demands on workers under normal conditions. These demands are amplified significantly when workarounds are needed, when problems occur, and when time is short. Howell and Cooke (1989) found that advances in technology and machine intelligence had in fact augmented, not lowered, the cognitive demands on humans. Basically what has been left for humans are the complex aspects of work such as tasks demanding judgment, assessment, diagnostic power, decision making, and the ability to plan and to anticipate future states. Also, the complexity of the work can boost cognitive demands such as the number of different factors to track, their diversity, and their level of interaction. Workers struggle to figure out how things interact and how outputs are produced from inputs.

Braz et al. (2007) have demonstrated that users have to manage complexity when authenticating to a Multifunction Teller Machine (MTM) using a mobile phone which is equipped with a special chip that communicates with the MTM. In addition, users will still be required to authenticate to the system by entering a PIN. Basically, the user scenario here involves a user making her monthly mortgage payment. In this case, the user has to deal with different services offered through different types of communication channels such as MTM, the Web, and Wireless Networks. Although it might be considered a convenient service when one does not have physical access to an MTM, it does place the burden on the users with regards to the coordination of the MTM and mobile phone.

User authentication systems are ultimately used by people, so their ease of use, understandability, satisfaction, and implicit cognitive dimensions must be addressed as well. The cognitive dimensions can be essentially considered as the interaction occurring between users and (security) user authentication mechanisms (e.g. logging into a system, interacting with an authentication token or smart card, and so on). When a user performs a task, that is, activities that are undertaken to achieve a goal (e.g. a user logs onto Bank of America online banking using a smartphone to access a checking account), some of these activities are considered *physical* (e.g. a user enters a password on the smartphone keyboard), while others are *cognitive* ones (e.g. user retrieves a password stored in his memory). In particular, *interface* is viewed by the authors either as a software component (e.g. a login webpage) or a hardware component (e.g. an authentication token, a smartphone, etc.) through which the interaction and information travel between the "interface" (software or hardware) and the user (Maffezzini, 2006).

Most usability inspection techniques do not overtly take into account users' thinking, "even though psychology-based inspection techniques supplied key insights into how thinking shapes interaction" (Hornbæk & Frøkjær, 2004). Evidence shows that the well-known Knowledge-Based Authentication does not take into account how people think (Adam & Sasse, 1999). Also, empirical research (Zurko & Simon, 1996), (Whitten & Tygar, 1998), and (Chiasson & Biddle, 2007) has shown that cognitive dimensions definitely influence the usability of security mechanisms under which user authentication methods are included. Researchers argue that security concepts used in security mechanisms are not easily understood by many users. Hence security designers should place additional effort into understanding the users' mental model and be certain to employ concepts the users can recognize.

For example, in a typical authentication task, Alice tries to log into a corporate computer system with a user ID and password. The activities that are undertaken to achieve this goal can be considered physical (e.g. Alice types in a password on a desktop keyboard) or cognitive (e.g. Alice tries to retrieve a strong password such as Gyz!152# stored in her memory which results in a huge demand on her memory). A strong password must be enforced, given that it makes the attacker's job much harder in guessing predictable passwords. This is not an easy task given that the cognitive capacity of a user to remember a password is quite limited (Sasse et al., 2001). But what is a strong password policy? It must contain at least eight characters, one uppercase alphabet (A-Z), one lowercase alphabet (a-z), one Arabic numeral (0–9), one non-alphanumeric character excluding "@!%^*&#(-". At this moment the system has blocked her account due to three unsuccessful attempts to log into the system. An authentication system should *a priori* promote strong passwords which account for security while still preserving memorability, which in turn accounts for usability. Another example is the poorly understood, overly complex, and hard to use Public Key authentication method according to usability evaluations (Whitten and Tygar, 1999) and (Williams and Voigt, 2004). For example, when Pretty Good Privacy (PGP) is in the process of encrypting or signing a file, it gives the user a status message indicating that it is presently "encoding." However, a better term would be "encrypting" or "signing", given that employing terms that overtly match the operations being executed helps to create an understandable mental model for the user.

These facets of understanding how users cope (or not) with different types of user authentication methods explain our interest in studying its cognitive dimensions in order to give a cognitive ergonomics account of user authentication design using Cognitive Task Analysis (CTA) (Hollnagel, 2003). CTA describes the physical tasks and cognitive plans required of a user to accomplish a particular work goal. The GOMS model was the method chosen to perform the CTA by the authors. The GOMS CTA method provides the capability of identifying, describing, and detailing the cognitive thought processes involved in the preparation and successful execution of user authentication procedures.

6.2.1 Introduce Cognitive Ergonomics

HCI involves systems comprised of people, computers, and their interactions. Cognitive Ergonomics (CE), however, is concerned with the mental aspects of the interaction, that is, the analysis of cognitive processes, such as perception, memory, reasoning, and motor response, required of operators in modern industries. CE is also concerned with developing specifications of the knowledge required by the human to interact with the computer to perform work effectively. These specifications are implementable as an interaction. It places particular emphasis on the analysis of cognitive processes (e.g. diagnosis, decision making, and planning) required of operators in modern industries.

CE aims to enhance the performance of cognitive tasks through several interventions including:

- user-centered design of human-machine interaction and human–computer interaction (HCI),
- design of information technology systems that support cognitive tasks (e.g. cognitive artifacts),
- development of training programs, and
- work redesign to manage cognitive workload and increase human reliability.

A major factor when designing security applications that must be taken into consideration by designers is that for the vast majority of users, security is an "enabling task" to one or more "production tasks" (e.g. access a database, shop online, etc.). Such an "enabling task" is perceived as an obstacle. In addition to that, cognitive demands required by authentication tasks are becoming increasingly complex. To reduce management and support costs, organizations are placing more and more of the burden of authentication on the user, forcing them to perform – at the enterprise's discretion – lifecycle-management tasks such as token requests and activation, password replacement, certificate renewal, etc.

The cognitive demands required by an assessment item are related to the number and strength of connections of concepts and procedures that a user needs to make in order to generate a response, in this particular book topic, when authenticating to a system (the assessment item). The cognitive processes are typically comprised of recall and recognition (e.g. face recognition authentication) and identification and classification [e.g. KBA such as SiteKey (BankofAmerica, 2009): first you recognize a unique image you chose and image title you created to accompany your image, then you mentally group image and title, carrying out in this way collection and comparison].

Traditionally CE has used the "human information-processing" model of cognition (Wickens, 1992), which models human cognition through a computer metaphor.

Central in CE is the notion of *domain*: Domain is the larger environment in which the work-system must operate, and presents both constraints and opportunities for the work-system. The domain influences the approach followed, as the degree of coupling among its constituents, the level of top-down causality, and the degree of human intentionality in decision making shapes the validity of the models used (Dowell & Long, 1998) (Figure 6.2).

CE also studies the competencies and limitations of the workers in their interactions with the work-system in general (e.g. attention, perception errors, strategies, cognitive workload), and in particular the cognitive artifacts they use to achieve their goals as well as their cooperation with other actors. The work-system is the system of agents interacting with each other to perform work by intentionally changing the states of domain objects.

CE is especially important in the design of complex, high-tech, or automated systems. A poorly designed mobile phone user interface may not cause an accident, but it may well cause great frustration on the part of the consumer and result in a business failure. However, a poor interface design on industrial automated

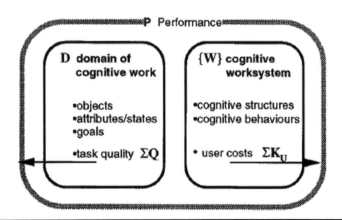

Figure 6.2   Work-system and a domain (Dowell and Long, 1998).

equipment may result in decreased production and quality, or even a life-threatening accident.

6.2.1.1 Methods

Methods used by Cognitive Ergonomics are those whose product yields specifications of knowledge. That is, the representations and processes required to support the user's behavior, such that, in interacting, the user and the computer achieve the desired performance. A wide range of methods are used in CE, but the most common methods are the Hierarchical Task Analysis (Kirwan et al., 1992) and Cognitive Task Analysis (CTA) (Hollnagel, 2003) and (Kieras, 1996). The Section 5.4 Develop the Natural GOMS Language (NGOMSL) describes the adopted GOMS model (Kieras, 1996) as the basis for the development of the CTA.

6.2.1.2 The Cognitive Approach

The cognitive approach allows the analyst to gather information and understand operation up to the thought process level. It allows a deeper understanding of the business problems or needs. This thorough understating can then be translated into better decision making. Overall, the cognitive approach is comprised of the rigorous practice of gathering information, human information processing, analysis, business modeling, and simulation as detailed in the succeeding paragraphs.

- *Gathering information*: Instead of relying only on meetings, surveys, or internal documentation, information is gathered in the field with "thinking out loud" techniques while people are performing their tasks. This ensures a deeper understanding of the current situation. Even if a process is totally changed, gathering information with the cognitive approach exceeds the risks of not doing so.
- *Information processing*: To understand the thinking process, goals, and knowledge, a CTA is performed which is undertaken in this book by the GOMS model. Cognitive goals, sub-goals, and methods are then described hierarchically. Methods are extracted with "how" questions, and goals are extracted with "why" questions. At the end of the process, management will have a deep understanding of the operations, problems, and strategies. This ensures an effective way to optimize any process. CTA serves also as input for defining the requirements of an information system.
- *Process Modeling*: The business is modeled as a hierarchy of systems and processes. The highest level is the mission, followed by generic functions, specific functions, and ultimately, at the most detailed level, the structural elements. The gathering techniques, along with the CTA, ensure that the business model will be grounded in reality. This provides a complete and exact picture of the operation to management.

- *Simulation*: Before executing a plan, each risk is analyzed and addressed by simulation and calculation. For example, in Information Technology (IT), user acceptance is often the major risk. Simulating the user interface prior to writing any line of code ensures that both users and business needs are met first. The application of the cognitive approach helps an organization to learn and translate that learning into rapid action, which is the vital competitive advantage.

6.2.2 Identify and Explain the Main Cognitive Areas of Focus Relating to User Authentication

This section addresses the main cognitive processes specifically related to Human–Computer Interaction (HCI) and user authentication. As already mentioned, the cognitive dimensions can be considered as the interaction occurring between users, and security (authentication) mechanisms typically embedded in Web, mobile, and/or network applications.

6.2.2.1 Perception

Perception is our awareness and understanding of the elements and objects of our environment through the physical sensation of our various senses, including sight, sound, smell, and so forth. Each sense organ is part of a sensory system which receives sensory inputs and transmits sensory information to the brain. Perception is influenced in part by experience. We classify stimuli based on models stored in our memories, and in this way achieve understanding. Basically, we tend to match objects or sensations perceived to things we already know. Other perceptual characteristics that are relevant to the user authentication subject matter include:

- *Proximity*: Our eyes and mind see objects as belonging together if they are near each other in space.
- *Similarity*: Our eyes and mind see objects as belonging together if they share a common visual property, such as color, size, shape, brightness, or orientation.
- *Matching patterns*: We respond in the same way to the same shape in different sizes. For example, the letters of the alphabet have the same meaning, regardless of physical size.

According to Dowell and Long (1998), the user's cognitive behaviors are the processing of representations. So, perception is a process whereby a representation of the domain, often mediated by tools, is created. Neisser (1964) emphasizes that human experience depends on the stored mental schema, which guide exploring behavior and the perception of external contexts.

There are numerous theoretical accounts of perception, which can, in general, be divided into two groups: Bottom-up processing and Top-down processing.

■ *Bottom-up processing* is also known as data-driven processing because perception begins with the stimulus itself. Processing is carried out in one direction from the retina to the visual cortex, with each successive stage in the visual pathway carrying out ever more complex analyses of the input.

■ *Top-down processing* refers to the use of contextual information in pattern recognition. For example, understanding difficult handwriting is easier when reading complete sentences than when reading single, isolated words. This is because the meaning of the surrounding words provides a context to aid understanding.

6.2.2.2 Memory

Memory is just one of many phenomena that show the brain's complexity. On a basic level, memory is the capacity for storing and retrieving information, but memories are not simply recorded and neatly stored. Our memories are selected, constructed, and edited not just by us but by the world around us. We have an amazing, unlimited capacity for memory, but our memories are also faulty, full of holes and distortions, and vulnerable due to unreliable data retrieval systems.

Memory can take different forms, and when we store a memory, we store information. However, what that information is and how long we retain it determines what type of memory it is. The biggest categories of memory are short-term memory (or working memory) and long-term memory, based on the amount of time the memory is stored. Both can weaken due to age, or a variety of other reasons and clinical conditions that affect memory (Figure 6.3). Memory can store, identify, and classify detailed sensory images, facts about the world, tasks mechanics, and experiences. Three processes are involved in memory: *encoding*, *storage*, and *retrieval*. All three of these processes determine whether something is remembered or forgotten.

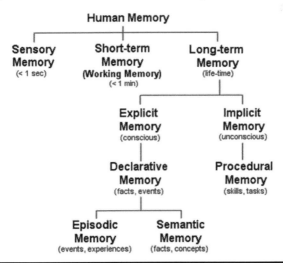

Figure 6.3 Types of human memory (Mastin, L., 2010).

■ *Encoding*: Processing information into memory is called *encoding*. People automatically encode some types of information without being aware of it. For example, most people can probably recall where they ate at lunch yesterday, even though they didn't try to remember this information. However, other types of information become encoded only if people pay attention to it. College students will probably not remember all the material in their textbooks unless they pay close attention while they are reading. There are several different ways of encoding verbal information:

   – Structural encoding focuses on what words look like. For instance, one might note whether words are long or short, in uppercase or lowercase, or handwritten or typed (e.g. a strong password Blitz4three$).

   – Phonemic encoding focuses on how words sound.

   – Semantic encoding focuses on the meaning of words. This requires a deeper level of processing than structural or phonemic encoding and usually results in better memory.

6.2.2.3 Storage

After information enters the brain, it has to be stored or maintained. To describe the process of storage, many psychologists use the three-stage model of memory proposed by Atkinson and Shiffrin (1968) as shown in Figure 6.4. According to this model, information is stored sequentially in three memory systems: Sensory Memory, Short-Term Memory (STM), and Long-Term Memory (LTM).

The following describes a detailed explanation of these three types of memory:

■ *Sensory memory*, or sensory register, notes or registers sensory stimuli as they are experienced. It consists of representations of the outside world as experienced through the senses such as touch, sight, or smell. It holds information for approximately one to two seconds. If, for instance, you glance at the ocean and turn away, the image of the ocean will be lost in one to two seconds unless the image is quickly transferred into the STM. The contents of sensory memory are constantly changing as new stimuli are perceived. Information that does not fade from sensory memory enters STM.

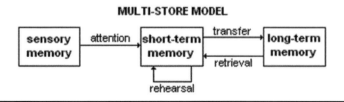

Figure 6.4   The three-stage processing model of memory (Atkinson and Shiffrin, 1968).

▪ *Working Memory (or Short-Term Memory)*: WM is thought to process information by actively repeating, grouping, and summarizing it to aid its storage in LTM. Information is thought to last within STM for only a short period of time before it is either passed into LTM or discarded. For information to be transferred into LTM, it must be rehearsed or repeated. WM is generally considered to have limited capacity. The earliest quantification of the capacity limit related with WM was the 7 ± 2 (i.e. the magical Number Seven, Plus or Minus Two) "chunks" of information rule introduced by Miller (1956). He noticed that the memory span of young adults was around seven elements, called chunks, regardless of whether the elements were digits, letters, words, or other units. The 7 ± 2 is that the capacity of STM is seven plus or minus two pieces of information (some people can hold five or six items, while others can hold eight or nine). According to Mathy and Feldman (2012), recent studies suggest that the number of items an individual can process using STM is closer to four, plus or minus one (Cowan, 2001). Irrespective of the exact number, these studies aim to elucidate that rather than trying to process information as a single, monolithic piece of information, breaking it down into smaller "chunks" can assist an individual to process and convert information from STM into LTM. Cowan argues that Miller's number was, in fact, meant more "as a rough estimate and a rhetorical device than as a real capacity limit. Others have since suggested that there is a more precise capacity limit, but that it is only three to five chunks." In his study, Cowan put together a broad range of data on capacity limits, asserting that the "smaller capacity limit is real".

In general, WM can hold five to nine units of information for between twenty seconds to one minute in length. It holds information for as long as it is actively thought about, or until new information basically forces it out. Unless we repeat the information and purposely try to retain it, most (or all) of it will be lost. A good example of this process can be seen when you look up a new phone number and repeat it to yourself as you dial it. After dialing it, within a few seconds you will usually forget it. However, if you do this repeatedly (repetition or rehearsal), as in the case of a friend with a new phone number, it will in the end up entering LTM. These "units" of information can represent single pieces of information, such as an individual's name, or the units can be single pieces of information that represent a number of different pieces of information, as in the last name of a family representing all of the family's members. The process of using a single item to represent a number of items is called chunking, and researchers have found that WM's information holding capacity can be significantly improved with this process. It seems that there are many factors that determine what information enters LTM, two of the strongest being repetition and intense emotion. If something is repeated often enough, such as multiplication tables, it will enter LTM. And it is hard to forget intensely emotional experiences, such as being involved in a serious car accident or falling in love. A subcategory of declarative memory is the ability to recognize previously encountered events, objects, or people.

■ *Long-Term Memory (LTM)*: LTM has been the focus of most research and theory on the memory system. It holds all the information that has managed to pass through the sensory and STM systems. In contrast to both of those systems, LTM is thought to be able to hold potentially unlimited amounts of information for an indefinite period of time, possibly for a lifetime. There is a structure for storing a representation of the knowledge we accumulate over time in the LTM. It is thought to hold all of the memories of our life, as well as our knowledge of the world in general. Information entering the LTM is assumed to be permanent. In LTM, one might find memories as diverse as the first person you ever had a crush on, knowledge of how to ride a bike, or cook scrambled eggs, or learn a second language. It is also where mental models are stored. Whereas STM generally holds between five and nine items, scientists say there are no limits on the capacity of LTM given that people have associations for those memories. That is why, for example, people have a natural inclination to choose passwords based on familiar things such as children's birthdays or favorite sports teams rather than incomprehensible strings like **3B#$Ir** or **7*$3fg**.

LTM stores and operates quite differently depending on the type of information involved. One of the most influential theoretical divisions of LTM is the division between Declarative Memory (episodic memory and semantic memory) and Procedural memory as described below:

– *Declarative memory (DM)* is the recall of factual information such as dates, words, faces, events, and concepts. It is so called because it refers to memories that can be consciously discussed, or declared. It applies to standard textbook learning and knowledge, as well as, for instance, remembering the capital of Germany, the rules for playing ice hockey, and what happened in the last episode of a streaming series as each of these examples involve declarative memory. DM is often considered to be explicit because it involves conscious, intentional remembering. It is subject to forgetting, but if regularly accessed, memories can last indefinitely. DMs are best established by using active recall combined with mnemonic techniques and spaced repetition. A mnemonic device is a memory and/or learning aid. Mnemonics are frequently verbal, something such as a very short poem or a special word used to help a person remember something, particularly a list, but they may also be visual, kinaesthetic, or auditory (Tulving and Schacter, 1990). Visual Recognition Memory is the ability to recognize elements in the surrounding environment, such as faces or places, as well as the ability to learn about and orient ourselves within that environment, both of which are crucial to our functioning in the world. We need to recognize individuals, such as family members, and to be able to navigate from one place to another. Thus neural systems have evolved to interpret incoming sensory information, with neurons that are capable of distinguishing novel and familiar visual elements. Systems based on recognition of visual items for authentication have been receiving much attention. For example, both Déjà Vu (Dhamija and Perrig, 2000) and Passfaces™ (Passfaces, 2009) present users with panels of images, from which they have to recognize and select their pass images. The most significant difference between the two systems involves the content of the images: Déjà Vu employs randomly generated art, while Passfaces uses photographs of strangers' faces in an attempt to exploit people's ability to process and remember faces. These systems have performed well in laboratory-style tests, producing recall rates of up to 80% even after up to three months of non-use (Valentine, 1998).

Declarative memory is divided into two types:

- *Semantic Memory* is the recall of general facts (e.g. remembering the rules for playing football).
- *Episodic Memory* is the recall of personal facts (e.g. remembering the city you were born in and the date). Episodic memory is the conscious recollection or recall of specific experiences from a person's life. These memories often include the time and place of the experience, as well as a representation of the role the individual who is remembering played in it. Episodic Memories seem to be more affected by the passage of time than are procedural or semantic memories, such that if the event is not recalled and thought of relatively often, details of the event, if not the event itself, seem to fade or be forgotten over time. Two specific types of mental representations hypothesized to be used by the semantic memory system in order to organize information are schemas (Anderson, 1977) and categories. Schemas are ordered frameworks or outlines of world knowledge that help us organize and interpret new information. They are like maps or blueprints into which new related information will be fitted. Knowledge of your home town or city, with its streets, various buildings, and neighborhoods, is an example of a schema. Schemas also help people to reconstruct, or try to remember, information that may have been forgotten. For example, if a friend brings up a time when you both went out to eat dinner a few months ago and you don't remember it clearly, you might ask for more information, and then use your schema for the usual sequence of events in eating out to try to remember or reconstruct what happened. Categories are another representational form of thought used by semantic memory to organize information. Categories are sets of objects, experiences, or ideas that are grouped together because they are similar to one another in some respect. For example, apartments, houses, huts, and igloos, might be grouped

under the category of dwellings. Like schemas, categories help us make sense of, and organize, the countless aspects of the world.

– *Procedural Memory* (*PM*) pertains to the storage of skills and procedures. This type of memory has also been referred to as "tacit knowledge" or "implicit knowledge". PM is involved in tasks such as remembering how to play squash or how to ride a bike. This is "know how" memory; it often can only be expressed by performing the specific skill, and people have problems verbalizing what they are doing and why. Procedural memory is therefore very important in human motor performance. An important feature of procedural memory is that it tends to persist; it's resistant to change – it can be useful since you don't want to have to keep re-learning behaviors. But this also means that you can't change a procedure, unless and until you pay attention to how and when it operates. An interesting characteristic of PM is that procedural patterns take a while to unlearn. Consider this scenario: You like to play tennis and have played for years. You decide to take some lessons. The instructor shows you how to swing the racquet more effectively. But you soon discover that you just can't tell yourself to swing it differently. The old pathways interfere with the new ones. It's hard to interrupt a well-established procedure. In fact, those original neural pathways, though weakened, will always be there, we currently have no reason to think that they will deteriorate. Under conditions resembling the initial circumstances in which they were laid down, they may even be reactivated. However, the new regulated pathways will eventually override the old ones.

■ *Priming* refers to an increased sensitivity to certain stimuli due to prior experience. Because priming is believed to occur outside of conscious awareness, it is different from memory that relies on the direct retrieval of information. Direct retrieval utilizes explicit memory, while priming relies on implicit memory. Priming can be conceptual or perceptual. *Conceptual priming* occurs where related ideas are used to prime the response, and is enhanced by semantic tasks. For example, *table* will show priming effects on *chair*, because *table* and *chair* belong to the same category. *Perceptual priming* is based on the form of the stimulus and is enhanced by the match between the early and later stimuli, for example, where a partial picture is completed based on a picture seen earlier.

6.2.2.4 Information Retrieval

There are two types of information retrieval: recall and recognition. In recognition, the presentation of the information provides the knowledge that the information has been seen before. Recognition is of lesser complexity, as the information is provided as a cue. However, recall can be assisted by the provision of retrieval cues, which enable the subject to quickly access the information in memory.

■ Recall

One of the most critical HCI principles is to avoid unaided recall wherever possible, since it is known to place a considerable burden on users' cognitive load and overall ability to perform. There are authentication mechanisms that use cued recall and recognition, for example:

- *Mutual authentication*: SiteKey (BankofAmerica, 2009) was a web-based security system that provided one type of mutual authentication between end users and websites. Its primary purpose was to deter phishing.
- *Cognitive passwords* involve a series of questions (e.g. security questions) about the user's personal preferences and history: After a predefined number of successful answers, the user is considered to have passed authentication.
- *Associative passwords* make use of users selected cues and responses. Its concept extends to a prespecified set of questions and answers that users would be expected to know and could easily recall. For example, users may associate passwords profiles around the rock band "The Beatles". In this case, cues may include "abbey", "John", "yellow", and "George" and have responses of "road", "Lennon", "submarine", and "Harrison", respectively.
- *Pass sentence*: It refers primarily to an unassisted recall mechanism in the first place-however, if the user does not get the secret completely right, the user is prompted with questions about the *pass sentence*, and when the user answers enough questions correctly, login is allowed.

■ Recognition

Recognition is one of the three basic memory tasks. It involves identifying objects or events that have been encountered before. Recognition is a process that occurs in thinking when some event, process, pattern, or object recurs; it involves knowing or feeling that someone or something present has been encountered before. Coming from the base cognition, recognition has various uses in different fields of study, and has generally been accepted as referring to the process of awareness or thought.

In psychology, cognition is used for information processing view of a person's psychological functions. This takes place as we process stimuli in relation to previous memories and experiences; also, we make connections between the current stimuli and our memories. Thus, in order for something to be recognized, it must be familiar. This recurrence allows the recognizer to more properly react. Hence recognition is a survival mechanism. Humans and animals will recognize certain foods, which are poisonous through taste, as they have tasted them before. This also works for sounds and alarms, which we are trained to react to, such as fire alarms.

Without recognition, we would go through life undergoing everything without learning from the past. Experiences would be pointless, as they would not be remembered. Recognition is the easiest of the memory tasks, and that is why multiple-choice tests are often considered easier than other tests. In multiple-choice tests, you only need to recognize the right answer. You do not have to come up with the answer on your own. Recognition uses the memories we have in place to help with the current situation. When the recognizer has correctly responded, this is a measure of understanding.

6.2.2.5 Password Memorability Issues

The user characteristic that has a major impact on password design is memorability. The Password authentication mechanism is in fact a huge component of the study of usable security in user authentication. As a Knowledge-Based Authentication (KBA) mechanism, it requires users to memorize items and recall them when accessing a protected computer system. Asking users to recall a single password and user ID for one system may seem reasonable, but with the proliferation of passwords, users are increasingly unable to cope. Research on human memory is extensive, but according to Sasse et al. (2001), the most important issues related to passwords can be summarized as follows:

- the capacity of working memory is limited;
- memory decays over time, meaning that people may not recall an item, or may not recall it 100% correctly;
- recognition of a familiar item is easier than unaided recall;
- frequently recalled items are easier to remember than infrequently used ones, and retrieval of very frequently recalled items becomes "automatic";
- people cannot "forget on demand", items will linger in memory even when they are no longer needed;
- items that are meaningful (such as words) are easier to recall than non-meaningful ones (sequences of letters and numbers that have no particular meaning);
- distinct items can be associated with each other to facilitate recall, yet similar items compete against each other on recall.

Research conducted by Sasse et al. (2001) regarding login failure (i.e. users forgetting passwords) found that the login usually failed because they recalled the password partly, but not 100% correctly and they recalled a different password from the one required (e.g. a previously used password for the same machine, or a password for a different machine). This demonstrates the basic memory mechanisms (described above) in action: Items decay in memory unless they are frequently recalled, and recall of similar items causes interference. The likelihood of 100% correct recall of infrequently used items is extremely low. This means that a password mechanism that demands 100% accurate recall every time is a bad match for infrequently used systems; that the results for 6-digit PINs are even worse confirms the importance of password content. It also indicates that a token-PIN combination, frequently proclaimed as a more usable substitute for passwords, is likely to cause more problems with infrequently used systems than a standard password. In second Study (the analysis of the password resets), the author found that 91.7% of resets were caused by "normal users" (i.e. more than 90% of users cannot cope with the password mechanism in the way they were expected to, which is a negative result in terms of the usability of password mechanisms).

As of this writing, the introduction of enterprise and consumer password management tools (i.e., roboform.com, keepersecurity.com, etc.) aims to assist users coping with the most employed KBA method, that is, passwords - these tools take the burden of remembering each login off the user. However, the vast majority of novice computer users typically do not adopt a password manager due to their little or no knowledge of computers and their uses, or even the motivation to do so. Finally, while password manager apps and browser plugins bring also their own security flaws (i.e., it can also be hacked), they are still a relevant tool in the battle to keep users information away from prying eyes.

■ Password Policies

The growing number of systems with which users have to interact and login to creates memory problems as already mentioned. The problem is often exacerbated by password policies creation and protection used by companies. The purpose of a password policy is to make sure that all company resources and data receive adequate password protection. The policy covers all employees who are responsible for one or more accounts or have access to any protected resource that requires a password.

■ Password creation:
- All passwords should be reasonably complex and difficult for unauthorized people to guess. Employees should choose passwords that are at least eight characters long and contain a combination of upper- and lowercase letters, numbers, and punctuation marks and other special characters. These requirements will be enforced with software when possible.
- In addition to meeting those requirements, employees should also use common sense when choosing passwords. They must avoid basic combinations that are easy to crack. For instance, choices like "password", "password1" and "Pa$$w0rd" are equally bad from a security perspective.
- A password should be unique, with meaning only to the employee who chooses it. That means dictionary words, common phrases, and even names should be avoided. One recommended method to choosing a strong password that is still easy to remember: Pick a phrase, take its initials, and replace some of those letters with numbers and other characters and mix up the capitalization. For example, the phrase "This may be one way to remember" can become "TmB0WTr!".

- Employees must choose unique passwords for all of their company accounts, and may not use a password that they are already using for a personal account.
- All passwords must be changed regularly, with the frequency varying based on the sensitivity of the account in question. For example, a common frequency is 90 days for a protected computer system such as Windows.
- If the security of a password is in doubt – for example, if it appears that an unauthorized person has logged in to the account – the password must be changed immediately.
- Default passwords – such as those created for new employees when they start a new job or those that protect new systems when they are initially set up – must be changed as quickly as possible.

■ **Protecting passwords:**

- Employees may never share their passwords with anyone else in the company, including co-workers, managers, administrative assistants, IT staff members, etc. Everyone who needs access to a system will be given their own unique password.
- Employees may never share their passwords with any outside parties, including those claiming to be representatives of a business partner with a legitimate need to access a system.
- Employees should take steps to avoid phishing scams and other attempts by hackers to steal passwords and other sensitive information. All employees will receive training on how to recognize these attacks.
- Employees must refrain from writing passwords down and keeping them at their workstations. See above for advice on creating memorable but secure passwords.
- Employees may use enterprise password management solutions approved by the organization's Information Technology (IT) department.

■ **Variability of Password Systems**

In most password systems, there is a great variability of user IDs and passwords across different systems (e.g. password policies for UNIX take up to 12 characters, Office 365 up to 16, etc.). Some systems have highly elaborate content restrictions, or more specifically, password security policies (e.g. a password must include upper/lowercase, at least six characters, alpha-numeric, and special characters), but these vary from system to system. The result is a huge demand on users' memories:

■ users not only have to remember passwords, but also the domain and user ID with which it is associated with (e.g., bns\s1232);

- users have to remember which password restrictions apply to which system;
- users have to remember whether they have changed a password on a particular system (particularly when there's no Single-Sign-On (SSO) system where users are required to create different credentials for accessing different systems), and what they have changed it to.

6.2.2.6 Mental Models

Mental models are representations of the function and/or structure of objects in peoples' minds. Designing something requires that you understand what the person wants to get done. Empathy with a person is distinct from studying how a person uses something. Empathy extends to knowing what the person wants to accomplish regardless of whether she has or is aware of the thing you are designing. You need to know the person's goals and what procedure and philosophy she follows to accomplish them. So mental models give you a deep understanding of people's motivations and thought processes, along with the emotional and philosophical landscape in which they are operating.

Norman (1988) defines a mental model as a system causality conveyance that is created by the user to reason about the system, anticipate its behavior, and explain why it reacts as it does, shown in Figure 6.5. At the same time, the designers create their own mental model for defining the system functioning (the conceptual model of the system) to simulate how the user will perceive it on the basis of the accessibility and the usability rules and principles. This is an attempt to clarify how the dialogue between user and technology in the interaction process is based on both the user's and the designers' mental models.

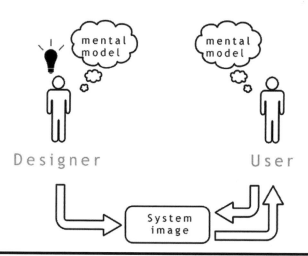

Figure 6.5   Mental Models (Norman, 1988).

Some characteristics of mental models are:

- They may be incorrect or incomplete.
- They can be "executed".
- They are analogical representations, or a combination of analogical and propositional representations.
- They are dynamically constructed when required.

There are two main types of mental models:

- Functional models: Good for everyday use.
- Structural models: Good for breakdown situations; difficult to acquire from usage experience only.

In short, computer systems should be designed in such a way that users can quickly acquire a good functional model of the system which is in accordance with their task model.

Chiasson and Biddle (2007) have found that in the usable security literature, and within their own studies, discussions invariably turn to the problem of mental models. User interfaces for security fall short of fostering useful mental models for users. One frequently cited explanation is that security is a complex issue and that users need more education in the area. The authors partially agree with this argument. Not only is it short-sighted to assume that users will be adequately trained, but it is unrealistic to place such a burden on users.

The user interface should convey the information necessary for users to be able to easily predict and understand the consequences of their actions. This does not mean that users need to know the intricate details of how the system operates, but that they can form a reliable explanation in their minds that lets them interact successfully. The file managing metaphor is a good example: users understand that files can be placed in folders, opened, closed, thrown into the recycle bin, and so on. But at no point do users need to know the underlying details of file storage and manipulation, such as disk blocks, index tables, and disk head scheduling.

Security interfaces do not yet help users form such mental models, and in fact still assume that users will have an understanding of underlying security concepts. This places users in a vulnerable position. They lack the necessary knowledge, they must rely on inadequate interfaces to deduce what is happening, and they must make decisions that could potentially place them at risk. A wrong decision can give attackers valuable information or leave a user's system vulnerable. Alternatively, a wrong decision can also hinder a user's productivity because the security mechanisms now prohibit desired activities. It is not surprising that users prefer not to deal with security issues if they can circumvent them. Security interfaces must foster useful mental models. Therefore, researchers and designers must also be careful to accurately identify users' mental models when running usability studies so that an accurate and unbiased understanding of the usability of systems can be obtained. These are not easy tasks, but ones that must nevertheless be accomplished to achieve usable security.

6.2.3 Develop the Cognitive Model of User Authentication (CMUA)

One of the goals of the Cognitive Model of User Authentication (CMUA) is to demonstrate what and how cognitive processes, particularly attention and memory, are involved in user authentication tasks. CMUA also serves as the basis for the development of the Usable Security Symmetry inspection method. Furthermore, the GLEAN3 (GOMSL) (Kieras, 1999) and SOAR (State Operator and Result) (Laird, 2008) cognitive architectures which have been adapted for the development of the CMUA are briefly described in this section.

6.2.3.1 Why Use a Cognitive Architecture?

Before proceeding with CMUA, it is worth noting what a cognitive architecture is and what the approach to modeling is. Cognitive architectures are an approach to modeling behavior that presupposes that there are two components to behavior, the i) architecture and ii) knowledge. The architecture is comprised of cognitive mechanisms that are fixed across tasks and essentially fixed across individuals. These mechanisms usually comprise of some type of perception and motor output, some sort of central processor, some working memory or activation of declarative memory, and some way to store and apply procedures. These mechanisms are used to apply task knowledge to generate behavior. The aim is to outline a cognitive architecture that captures a selection of cognitive processes in an integrated manner, and, therefore, to provide integrated explanations of a broad array of data (Laird & Congdon, 2009) and (Ritter, 2004).

Cognitive architectures must embody strong hypotheses about the building blocks of cognition that are shared by all tasks, and how different types of knowledge are learned, encoded, and used, making a cognitive architecture a software implementation of a general theory of intelligence.

A basic cognitive architecture for which the **Goal** is to bring all existing **Knowledge** to allow selected actions (**Task Environment, Body**) to achieve goals is shown in Figure 6.6. Learning is implied in Knowledge.

There are several types of architectures that are or that could be used for evaluating interfaces and predicting task time and errors. These include descriptive architectures (Kieras, 1999), symbolic (Laird et al., 1987) and hybrid architectures, intelligent agent architectures, and connectionist architectures. CMUA looks at *how cognitive processes take place in a particular cognitive architecture*. To this end, the authors adapt and refine two recognized architectures for cognitive modeling, such as GLEAN3 (GOMSL) (Kieras, 1999) and SOAR (Laird, 2008), in order to

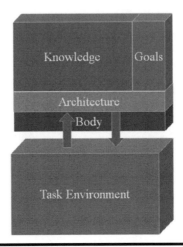

Figure 6.6   A basic cognitive architecture (Laird, 2009).

build the CMUA. The result is a descriptive architecture which helps in system design and offers a first view of how the human interaction process occurs in user authentication.

6.2.3.2 GLEAN3 (GOMS Language Evaluation and Analysis)

GLEAN3 is a newer computational form of NGOMSL, called GOMSL (Language) which is processed and executed by a GOMS model *simulation tool*, GLEAN3 (GOMS Language Evaluation and Analysis). GLEAN3 was inspired by the original GLEAN tool developed by Wood (1993) and implemented and elaborated by Kieras et al. (1995) and then again as summarized by Kieras (1999). Unlike the earlier versions, GLEAN3 is based on a comprehensive cognitive architecture, namely, a simplified version of the EPIC (Executive-Process/Interactive Control) architecture for simulating human cognition and performance (Kieras & Meyer, 1997). GOMS and GLEAN3 have been also used to identify likely sources of errors in user interfaces and model human error recovery (Wood, 2000) and also to model the performance of teams of humans who interact with speech (Santoro & Kieras, 2001).

GLEAN3 is a computationally realized version of the Model Human Processor (MHP), being based more on the EPIC (Executive-Process/Interactive Control) architecture (Kieras & Meyer, 1997) for human information processing that precisely accounts for the thorough timing of human perceptual, cognitive, and motor activity. EPIC provides a framework for constructing models of human system interaction that are precise and comprehensive as much as necessary to be useful for practical design purposes. EPIC depicts a state-of-the-art synthesis of results on human perceptual/motor performance, cognitive modeling techniques, and task analysis methodology, implemented in the form of computer simulation software. GLEAN has been used in numerous domains including military command

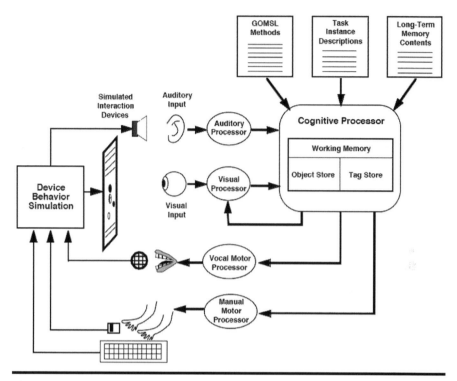

Figure 6.7   GLEAN3 cognitive architecture (Kieras, 1999).

and control, aircraft maintenance, and web applications. Current development is extending GLEAN to better support error analysis and error tolerant design (Kieras, 1996).

The GLEAN3 basic architectural structure is shown in Figure 6.7.

Human performance in a task is simulated by programming the cognitive processor with production rules organized as methods for accomplishing task goals. The EPIC model is then run in interaction with a simulation of the external system and performs the same task as the human operator would. The model generates events (e.g. eye movements, key strokes, vocal utterances) whose timing is accurately predictive of human performance.

6.2.3.3 SOAR (State Operator and Result) Cognitive Architecture

The SOAR architecture is a symbolic cognitive architecture, created by Laird et al. (1987) (Figure 6.8). It is both a vision of what cognition is and an implementation of that vision through a computer programming architecture for Artificial Intelligence (AI). SOAR has been broadly used by AI researchers to model diverse aspects of human behavior.

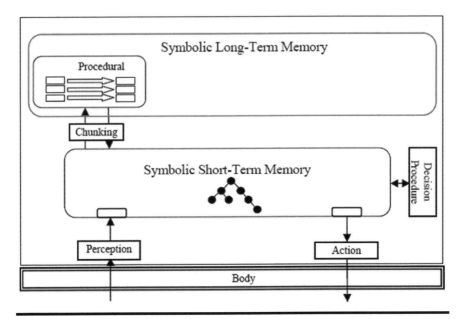

Figure 6.8 The SOAR (State Operator and Result) architecture (Laird et al., 1987).

The main goal of the SOAR project is to be capable of handling the full range of capabilities of an intelligent agent, from highly routine to exceedingly complex open-ended problems. To this end, it needs to be capable of generating representations and using appropriate forms of knowledge (i.e. procedural, declarative, episodic, and possibly iconic). SOAR should then deal with a collection of mechanisms of the mind. Also, underlying the SOAR is the vision that a symbolic system is required and enough for general intelligence. This is known as the *physical symbol system hypothesis* by Newell & Simon (1997) which states that "A physical symbol system has the necessary and sufficient means for general intelligent action."

SOAR consistently illustrates Short-Term Knowledge (STK) as a network of active symbols. Long-Term Knowledge (LTK) is a collection of condition-action rules. The conditions of each rule build up a pattern to match against the active symbols network. When a rule's condition matches, the rule runs by carrying out its actions. These actions might entail adding (or deleting) symbols in the STK structure.

To deal with complexity, SOAR includes a goal hierarchy, allowing successive decomposition of problems into component sub-problems. SOAR includes mechanisms to generate new goals automatically in response to a system's LTK and existing situation. SOAR depicts perceptual and conceptual knowledge consistently in STM; therefore, new actions flow from preceding actions and from changes in the external environment. The rule system also incorporates pattern-matching technology, allowing quick processing. Hence, SOAR is well suited for the development of intelligent systems that must produce actions in time analogous to human decision time.

Also, SOAR includes an automatic learning mechanism based on the psychological concept of chunking. SOAR learns new chunks by compiling sequences of actions that change STM in particular ways. New chunks fit consistently into a system's existing long-term rule set. Therefore, a SOAR system can incrementally learn new facts about the world, as well as more proficient representations of its original LTK.

6.2.3.4 Cognitive Model of User Authentication (CMUA) Cognitive Architecture

A cognitive model is a representation of some aspects of the user's understanding, knowledge, intentions, or processing. Building a model of how a user works allows us to foresee how s/he will interact with the user interface. One of the available modeling techniques, as already explained in previous sections, is the Model Human Processor (MHP). A basic model of human performance, it is intended to offer brute predictions of system behavior. MHP also considers humans as information processing systems such as a collection of memories and processors, as well as a set of principles ("principles of operation").

The CMUA is a cognitive model which has been built specifically to understand the cognitive processes involved in user authentication methods. Figure 6.9 shows the basic architectural structure of CMUA.

From the top, the CMUA consists of a **LTM**, which is encoded as production rules, and a STM, which is encoded as a symbolic graph structure so that objects are able to be represented with properties and relations. **Symbolic STM** holds the agent's evaluation of the current situation derived from perception and via retrieval of knowledge from its LTM. **Action** in an environment takes place through the creation of motor commands in a buffer in STM. The **Decision Procedure** selects operators and detects impasses. At the lowest level, CMUA's processing consists of matching and firing rules. Rules provide a flexible, context-dependent representation of knowledge, with their conditions matching the current situation and their actions retrieving information relevant to the current situation. In fact, many rule-based systems select a single rule to fire at a given time, and this serves as the venue of choice in the system – where one action is chosen instead of another action.

CMUA lets additional knowledge exert influence on a decision by introducing *operators* as the venue for choice and employing rules to propose, assess, and apply operators. Rules perform as an associative-memory that retrieves information pertinent to the present situation, so because of this, rules fire in parallel. In CMUA, there are rules that *propose* operators that generate a data structure in working memory representing the operator and an *acceptable preference* so that the operator is able to be taken into consideration for selection. There are also rules that *evaluate* operators and generate other categories of preferences that favor one operator

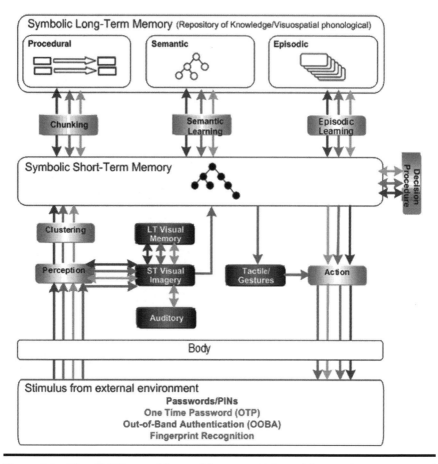

Figure 6.9   The CMUA cognitive architecture. (Adapted from (Kieras, 1999) and SOAR (Laird, 2008).)

to another or provide some sign of the usefulness of the operator for the present situation. Lastly, there are rules that *apply to* the operator by making changes to working memory that reflect the actions of the operator. These changes might be solely internal or start external actions in the environment. This approach supports a flexible representation of knowledge about operators. There can be many reasons for proposing, selecting, and/or applying an operator, some that are very specific and others that are quite general.

CMUA possesses a shared STM where knowledge from **Perception** and **LTM** are combined to offer an integrated representation of the current situation. It has an "incline" decision procedure that supports context-dependent reactive behavior, but in addition, supports automatic impasse-driven sub-goals and meta-reasoning. Chunking is CMUA's learning mechanism that converts the outcomes of problem-solving in sub-goals into rules compiling knowledge and behavior from deliberate to reactive (Laird, 2008).

CMUA shows the four most representative categories of user authentication methods as follows:

- Password/PINs (PPs).
- One-Time Password (OTP).
- Out-Of-Band-Authentication (OOBA).
- Fingerprint Recognition (FR).

Figure 6.9 depicts the user authentication tasks with their corresponding cognitive and motor process-flows gray-scale arrows as follows:

$$\rightarrow \text{PPs} \rightarrow \text{OTP} \rightarrow \text{OOBA} \rightarrow \text{FR}$$

6.2.3.4.1 CMUA Components

The following describes all CMUA components:

- **Stimulus from External Environment:** The external environment consists of PPs, OTP, OOBA, and FR user authentication tasks.
- **Body:** Users receive stimuli through their sense organs, the five senses (sensory information). Sensory memory corresponds approximately to the initial 200–500 milliseconds after an item is perceived.
- **Perception:** The Perception component is represented by the brain, which selects, organizes, and interprets sensory information.
- **Auditory:** The Auditory memory component accepts either speech or sound inputs and makes them available to the WM. It is the capability to remember what an individual has heard. It involves being able to take in information that is introduced to you, process that information, store it in STM, and then recall what you have heard. Mostly, it involves the task of attending to, listening, processing, storing, and recalling.
- **Visual Imagery:** The CMUA provides a set of processes and memories to support visual imagery, which includes depicted representations. The **ST Visual Imagery** module includes a STM where images are built and manipulated, a LTM that includes images that can be retrieved into the STM, processes that manipulate images in STM, and processes that generate symbolic structures from the visual images. Visual imagery is controlled by the symbolic system, which emits commands to build, manipulate, and inspect visual images. Additionally, visual imagery allows processing that is not possible with only symbolic reasoning, such as determining which letters in the alphabet are symmetric along the vertical axis (A, H, I, M, O, T, U, V, W, X, Y) (Laird, 2008).
- **Visual Memory:** Visual memory is an individual's capability to remember what s/he has seen. Typically, three types of visual memory are found: Photographic memory, iconic memory, and spatial memory. *Photographic memory* is the capability to recall images and/or objects in memory with great precision and in plentiful volume. *Iconic memory* is the sensory store for vision,

a type of ST visual memory. Experiments performed by Sperling (1960) offer an indication of a speedily decaying sensory trace, lasting roughly only 250 milliseconds after the offset of a display. *Spatial memory* can be considered a sub-category of visual memory because it relies on a cognitive map. Cognitive mapping is a sort of mental processing by which a person is able to acquire, code, store, recall, and decode information about the relative locations and attributes of events in their daily or symbolic spatial environment.

- **Tactile/Gestures**: A gesture is "an imprecise, context-dependent event that conveys the user's intentions" (Voyles et al., 1995). In this case, the gestures are force impulses – nudges – on the end-effector. Because gestures are context-dependent, state information (i.e. generally application specific) must be associated with each gesture. Using a linguistic analogy, the raw gestures form a gestural alphabet along with the state information. Gestural words are assembled from the raw gesture and its associated context by the gesture recognizers.
- **Clustering**: Classification is a basic human conceptual activity. For example, children gain very early knowledge of classifying objects in their environment and correlating the resulting classes with nouns in their language. "Cluster analysis" is a broad term used for a multiplicity of procedures that can be employed to create subtypes (i.e. classification). These procedures form "clusters" or groups of highly similar entities. More particularly, a clustering method is a multivariate statistical procedure that starts with a data set including information about a sample of entities and attempts to rearrange these entities into somewhat homogeneous groups. The **Clustering** component detects statistical regularities in the flow of experiences and automatically and dynamically creates new symbolic structures that represent those regularities, providing a mechanism for automatically generating new symbols and thus concepts that can be used to classify perception (Laird, 2008). Those new symbolic structures enrich that state representation. Clustering is, in fact, sub-symbolic, where non-symbolic perceptual structures are amalgamated collectively to create symbols.
- **Symbolic Short-Term Memory:** Symbolic STM holds the agent's assessment of the current situation derived from perception and via retrieval of knowledge from its LTM. STM allows recall for a period of several seconds to a minute without rehearsal, and its capacity is also very restricted. According to (Miller, 1956) experiments, the store of STM was 7 ± 2 items, the Magical Number Seven, Plus or Minus Two.
- **Decision Procedure**: The decision procedure selects operators and detects impasses. It helps context-dependent reactive behavior, but also helps automatic impasse-driven sub-goals and meta-reasoning.
- **Chunking**: Chunking means arranging items into familiar, manageable units. Each chunk collects a number of parts of information from the environment into a particular unit. Chunking refers to an approach for making more proficient use of STM by recoding information. In general, Herbert

A. Simon* has used the term *chunk* to indicate LTM structures that can be employed as units of perception and meaning, and *chunking* as the learning mechanisms guiding the acquirement of these chunks.

■ **Semantic Learning**: Declarative knowledge can be separated into elements that are known (i.e. facts) and elements that are remembered (i.e. episodic experiences). Semantic Learning and memory provide the capability to stock up and retrieve declarative facts about the world, such as cars have wheels, eggplant is a vegetable, and pyramids are in Egypt. This capability increases the ability to create agents that reason and employ general knowledge about the world. Semantic Learning is built up from structures that take place in STM. A structure from Semantic Learning is retrieved by generating a cue in a particular buffer in STM. The cue is then employed to seek for the best partial match in semantic memory, which is then retrieved into STM.

■ **Episodic Learning**: Episodic memory is the type of memory that remembers events that are observed through experience (e.g. a snapshot from one's past experience). It includes specific instances of the structures that occur in STM at the same time, providing the capability to remember the context of past experiences as well as the temporal relationships between experiences (Nuxoll & Laird, 2004). An episode is retrieved by generating a cue in a particular buffer in STM. The cue is then employed to seek the best partial match in semantic memory, which is then retrieved into STM. The next episode can also be retrieved, providing the capability to replay an experience as a succession of retrieved episodes. Episodic memory is task-independent and therefore available for every problem, providing a memory of experience not available from other mechanisms.

■ **Symbolic Long-Term Memory**: Symbolic LTM contains images that can be retrieved into the STM, and which are encoded as production rules.

■ **Action**: Action in an environment takes place through the generation of motor commands in a buffer in STM.

6.2.3.4.2 CMUA Processing Cycle

■ **Input**. Users receive stimuli through their sense organs, the five senses. "Stimulus from the external environment" is represented by the user authentication tasks (sensory information).

■ **Perception** is carried out through different perceptual processors such as visual memory, visual imagery, tactile gestures, and auditory. In *visual*

---

* Herbert. A. Simon was an American psychologist whose research ranged across the fields of cognitive psychology, computer science, public administration, economics, management, philosophy of science, and sociology. He was a professor at Carnegie Mellon University, Pittsburgh, PA. With nearly a thousand frequently quoted publications, he is one of the most authoritative social scientists of the 20th century.

memory, SiteKey is an example where first, you recognize a unique image you chose and an image title "Have a nice day" that accompanies the chosen image. In *visual imagery*, an example is when a user visualizes a 4-digit-PIN like "2222" in his mind and it visually associates with swans. In *tactile gestures*, a user places his finger on a mobile device that reads the thumbprint. Finally, in *auditory*, a user gets a sound alert such as a beep on her smartphone warning that she just received a text message, which includes the token code that she must authenticate with when using OOBA.

- **Changes to perception** are processed, and clustering is undertaken if needed. Then, changes are sent to the Symbolic STM.
- **Chunking, Semantic Learning,** and/or **Episodic Learning** are undertaken if needed. *Chunking*, in a random 10-character password, a chunk is a symbol, and, as Miller (1956) claims, the majority of people cannot remember 10 random symbols. Users are probably more certain to remember a 10-character pass phrase comprised of 2 or 3 words or chunks, let's say memorizing a *passphrase* (i.e. a long password with inserted spaces such as "My voice is my password"). In *Semantic Learning*, an emphasis is given on the importance of the interactive and psychological situations in which learning occurs. Learning is identified with the acquisition of knowledge (e.g. the user is required to memorize a sequence of 4 images of the *same category* (enrollment): let's say that the category is racquet sports so this would be represented by the following sports: badminton, racquetball, squash, and tennis). Then when later she authenticates to the system, a series of images is presented from categories that she has pre-selected mixed with images from random categories. After that, she retrieves it from LTM by entering that sequence in a fixed order. Finally, in *Episodic Learning*, a change in behavior takes place as a result of an event (e.g. to change a password, users enter and confirm a new password that is at least eight characters long and which includes at least one number. Users cannot reuse any of their previous passwords).
- **Elaboration**. Rules compute entailments of STM. For example, a rule might test if the goal is to grasp a hardware token by considering the hardware token's distance, and the user's reach, and afterwards create a structure signifying whether the hardware token is within reach.
- **Operator Application**. The actions of an operator are carried out by rules that match the present situation and the present operator structure. Several rules can shoot in parallel and in sequence offering a means for encoding operator actions.
- **Output (Action)**. Any output commands are passed on to the motor system.

6.2.4 Define the Usable Security Principles and Develop a Cross-Cognitive Analysis

This section specifies the usable security principles targeted to user authentication which were defined based on the research in the most recognized principles of HCI (Nielsen, 1994a), usable security guidelines and principles presented in Chapter 3

[(Chiasson et al., 2007), (Jøsang et al., 2007b), (Chiasson et al., 2006), (Garfinkel, 2005), (Yee, 2005), (Whitten & Tygar, 1998), (Saltzer & Schroeder, 1975)], and each of the main categories of user authentication methods. This will serve as part of the basis for the development of the usable security symmetry inspection method.

■ **Password/Passphrases/PINs**

Principle	Correlation
Visibility	Users hardly see the password they type, or if they do, the password is hidden under asterisks. The interface cannot provide visual cues, reminders, lists of choices, or other aids; the system cannot display the typed password as it is since it would be an open door for eavesdropping and Social Engineering attacks. Even though some systems provide the common used "eye icon" for showing passwords, the eye icon which when clicked shows the password, it still requires users another interaction with the system which can be annoying.
Feedback (Error Handling)	Most systems only mention success or failure. If an error is made, the system should be able to detect the error and offer simple, comprehensible mechanisms for handling the error. However, if it gives us clues like "the password has to contain letters" we will be exposed to dictionary, eavesdropping, and Social Engineering attacks.
Consistency	This authentication method, in general, consistently presents the same layout and terminology (e.g. username/ password login windows). The prompts are well recognized by users (e.g. novice, intermediate, and expert computer security end-users) as a security mechanism for accessing a protected network system. Besides, the association of a typed password with asterisks is very common among users.
Compatibility	This authentication method demands a lot of a user's STM load. The human mind can recognize better than recall, and this authentication method fails to provide enough information to the user so that the user can take action without recalling a lot.
Simplicity	It is simple and straightforward from the human factors and ergonomics standpoints (e.g. body movements such as move the mouse to the text field). Some functionalities like "Help" and "Forgot Your Password?" are usually displayed in hidden links on a web page but they can be activated by clicking on them.

■ Challenge–Response Calculators (CRC)

Principle	Correlation
Visibility	Users hardly see the password and PIN they type, or if they do, the password and PIN are hidden under asterisks. The interface cannot provide visual cues, reminders, lists of choices, or other aids; the system cannot display the typed password or PIN given that it would be an open door for eavesdropping and Social Engineering attacks. CRCs should provide ease of use by simple commands, or better, the user's task to authenticate her/himself should become as automatic as possible, increasing its speed, efficiency, and usability.
Feedback (Error Handling)	The system only mentions success or failure related to passwords, PINs, and "challenges". If an error is made, the system should be able to detect the error and offer simple, comprehensible mechanisms for handling the error.
Consistency	The user has to navigate to multiple screens in order to complete a task. The authentication process is as follows: • User enters a PIN. • Server sends a "challenge" to the user. • User reads the "challenge" from the token display and enters it on the token keypad. • A *response* is calculated by the server. • User reads the response on the display and enters it on the computer terminal. • The responses calculated by the token and server are compared, and if they are equal, the user is successfully authenticated. The challenge–response security dialog can be very annoying, which can negatively affect the user's experience.
Compatibility	CRCs require several extra steps (e.g. PIN + challenge + password) and additional memory load for users which can be very time-consuming. A stronger password including other characters other than digits will demand an even larger CRC keypad, which is not desirable.
Simplicity	A CRC that a user has logged into should be shut off within a given timeframe if not used to prevent illicit users from gaining access while the token is left unattended. However, this can be very annoying for the user, who might have to login with the token several times. Besides, handling an electronic device in a proper manner can be quite a challenge for many users.

■ Public Key (PK) Authentication

Principle	*Correlation*
Visibility	The controls of the PK method are not obviously visible and intuitive, and their functions are not recognizable; it is not clear when something has been encrypted and when it has not (Whitten & Tygar, 1999).
Feedback (Error Handling)	Very few errors are revertible, even if they demand some time and effort to rebuild the intended state. There are numerous irreversible actions in the PK system, such as accidentally deleting the private key, publicizing a key or revoking a key, forgetting the passphrase, and finally, failing to back up the key rings.
Consistency	To encrypt or sign a file, PK presents users with a status message that indicates it is now "encoding"; it would better to say "encrypting" or "signing", since being able to see terms that unequivocally match the operation being executed helps to create a clear mental model for users.
Compatibility	Users need a sensorial perception of the PK method in order to allow them to interact with the system toward achieving their goals, which don't occur in this type of authentication method since the PK model is not clear to users. Additionally, users are unlikely to put too much effort into tasks for which they don't understand the need.
Simplicity	The PK system is difficult to learn and use, and displays too much information. A separate certificate is needed for each Certificate Authority for different services and certificate expiration. Using a computer, installing and importing a certificate to other applications might be difficult as it can only be used at the location where it is installed. To use a smart card, a reader is necessary, and users must carry the token with them. A crucial task, such as making a backup revocation certificate, is not easy to perform. All these factors are not so convenient from the users' perspective. Furthermore, PK models are too demanding for users as we can see in the authentication process as follows: 1. User supplies her/his certificate with the Certificate Authority, a signature, and the PK to the authentication system; 2. User proves possession of the secret private key (PRK) by presenting the file or inserting the smartcard into a reader; 3. The authentication system checks if the PK in the certificate corresponds to the private secret key. If they correspond, the user has proven possession of the secret key and is successfully authenticated.

■ Biometrics

Principle	Correlation
Visibility	Generally speaking, registration, identification, and authentication processes are very time-consuming.[a]
Feedback (Error Handling)	Most systems only mention success or failure, and the user might be erroneously rejected.
Consistency	In fingerprint recognition, the exact placement of the fingertip on the scanning surface is very important for reliable performance (i.e. elastic distortion from one sample to the next) and sometimes users don't place their fingertip correctly. With face recognition, the concept of recognition itself is well known by users, usually resulting in high user acceptance.
Compatibility	Much training and education are necessary to prevent false rejects. Many biometrics cannot handle variations of the users' characteristics (e.g. aging, illness, injury, etc.).
Simplicity	With the voice recognition method, for example, it might need numerous live samples.

[a] Alan E. Zuckerman, M.D., Kenneth A. Moon, M.D., and Kenneth Eaddy: 2007. Comparison of Fingerprint and Iris Biometric Authentication for Control of Digital Signatures. January 3rd, 2016 http://www.ncbi.nlm.nih.gov/pmc/articles/PMC2244356/pdf/procamiasymp00001-1249.pdf

Chapter 7

The Usable Security Protocol Methodology: Formulate

The previous chapter has dealt with the cognitive processes involved in user authentication and their application in a cognitive architecture – the Cognitive Model of User Authentication (CMUA). Now that we understand what the cognitive processes are and how they generate responses when users authenticate to a system, it is time to put those cognitive processes into operation in computing systems.

7.1 Formulate the Usable Security Symmetry

Authentication mechanisms in complex systems such as Multifunction Teller Machines (MTMs) for banking are characterized by their user interface (UI) components. Furthermore, there is a common but false belief that security is only related to the software functionality and can be designed independently from the software usability which is related to the user interface component (Seffah and Metzker, 2004). In fact, the meaning of what a UI is and how usability is defined are perhaps major underlying obstacles that explain such erroneous conceptions. Indeed, it gives the impression that the UI is a thin layer sitting on top of the "real" system and that usability can be conceived of independently from the other quality factors such as security.

The human–computer interaction (security) (HCISec) research community has constantly been reporting the poor usability of security systems and its consequences, vulnerabilities, and threats (Whitten & Tygar, 1999), (Stiegler et al., 2004), and (Saltzer & Schroeder, 1975). Also, a significant number of usability problems causing security failures were found in the Pretty Good Privacy (PGP) study (Whitten and Tygar, 1999), a public key encryption program mainly intended for email privacy and authentication. The authors agree with the idea fundamentally supported by Whitten & Tygar (1999) that there is a need for a comprehensive model of *usable security* more specifically for user authentication methods. This model should include both process- and product-related usability characteristics such as effectiveness, efficiency, satisfaction, security, and learnability.

As mentioned, usable security is defined as the study of how elements of information security and usability should be handled together both at the front end and at the back end, while take into consideration the associated resources and costs. Usable security is imperative from the *user's perspective* (e.g. authenticating appropriately in a computer system without circumventing the security policy), from the *developer's perspective* (e.g. success or breakdown of a token-provisioning application), and from the *management's perspective* (e.g. enforcing a strong password policy can be a major constraint to the usability of a system).

Modeling is always a goal-driven activity since every model has a purpose. Testing and validation need to be done with the purpose of the model in mind. To this end, the validation phase is based on a *design-driven demonstration* of the CMUA by using one of the most representative authentication methods used for user authentication: One-time password (OTP).

The development of an authentication method is subject to the following phases: Evaluation, Definition, Development, and Readiness, as shown in Figure 7.1.

In 1, the Authentication Method Requirements Document (AMRD) version 1.0 is developed which comprises of a list of authentication method requirements. Next, in 2, the Authentication Method System Requirements (AMSR), a UXD Spec 0.x document, and the AM schedule are completed. Then, in 3, the development phase proceeds by executing the UXD Spec version 1.0, prototyping, usability testing, and user acceptance testing (i.e. quality engineering); finally, in 4, the readiness phase releases the authentication method to manufacturing and/ or market.

7.1.1 Security as a Usability Characteristic

Researchers, as well as standard organizations, have provided an additional perspective on usability that refers to a specific usability characteristic, which is *security*. Figure 7.2 lists some of the standards where security is included within their usability model as follows:

▪ Information Technology Security Evaluation Criteria (ITSEC) IEC 300: It presents software as security-critical.

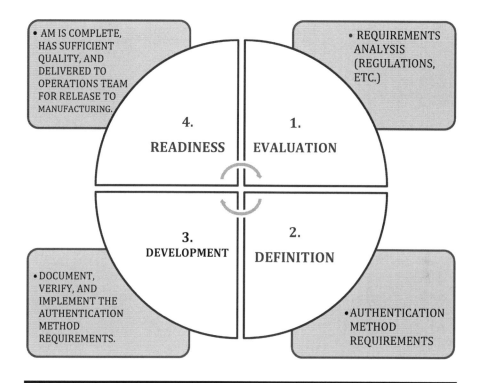

Figure 7.1 Authentication Method (AM) development lifecycle.

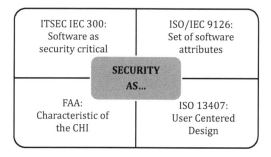

Figure 7.2 Security as a usability characteristic.

■ International Standards Organization (ISO)/IEC 13407: It describes human-centered design as a multidisciplinary activity incorporating human factors and ergonomic and technical knowledge with the objective of raising efficiency and effectiveness, improving human working conditions, and opposing possible unfavorable effects of use on human health, security, and performance.

- ISO/IEC 9126: It defines security, which is a sub-characteristic, as a set of software attributes that relate to its ability to prevent unauthorized access, whether accidental or deliberate, to programs and data.
- Federal Aviation Administration (FAA): Security is a characteristic of the CHI, which is particularly important in an industrial context.

These standards consider that good usability is a significant condition for human security in critical systems, such as medical apparatus or nuclear power stations. Within our model, the authors adopt this perspective of security.

The usable security community acknowledges that for a system to be secure, it has to be usable. This means that even the most secure system can fail if it is not used appropriately. In Whitten & Tygar's (1998) study, usability of security has different requirements than usability of IT in general. As already mentioned in the introduction of this book, it is broadly held that security and usability are two opposed goals in system design (Cranor & Garfinkel, 2005), (Jøsang et al., 2007b), and (Nielsen, 2000). However, there are several cases in which security and usability can be synergistically enhanced by reviewing the usable security approach. For example, improving the interface and changing the way users interact with the system (Yee, 2004), (Nielsen, 2000), and (Sasse et al., 2001). Additionally, Polaris (Stiegler et al., 2004) allows users to configure most applications so that they launch with only the rights they need in order to get the job done (i.e. the principle of least authority), thereby demonstrating that it is feasible to build systems that are more secure, more functional, and easier to use.

The users' roles as end users, IT administrators, developers, and security designers are one of the main concerns in the field of usable security. According to Zurko & Simon (1996), software developers working on UI design and evaluation lack user-centered design (UCD) tools. Tools are needed to support security designers in acquiring and sharing UCD and software engineering best practices. This is especially the case with user authentication. The usable security symmetry inspection method aims to help security designers to design, inspect, and evaluate the usability as well as the security aspects of user authentication mechanisms. From the authors' perspective, a security designer is an expert in computer security and possesses a reasonable understanding of the skills, mindset, and background of the users who are expected to perform an authentication task. The usable security symmetry integrates usable security earlier into the requirements and design phases of the user authentication methods development lifecycle.

7.1.2 Usability Factors and Usability Criteria Mapping

Generally speaking, the quality of a software product is specified by its internal and external capability to assist users in achieving their goals and the organization's goals, therefore improving productivity and human health. The ISO/IEC

9126-1:2001 standard is founded on a quality model for software products that consists of two parts: (1) external and internal quality, and (2) quality in use. In brief, internal quality refers to properties of the non-executable portion of a software product during its development, and metrics for internal quality, in general, refer to the quality of intermediary deliverables, for instance, the source code for a prototype version. External quality, in turn, refers to the behavior of the computer system of which the software product is a part.

To build the usability factors and usability criteria mapping, an adaptation of the Quality in Use Integrated Measurement (QUIM) (Seffah et al., 2006) hierarchical model has been used. "Quality in use is a kind of higher-order software quality construct, and it concerns whether a software product enables particular users to achieve specified goals with effectiveness, productivity, safety, and satisfaction in a specific context of use." QUIM adopts the viewpoint of most HCI standards that serve to foster quality in use in different factors: Usability is broken down into factors, then into criteria. For the purposes of this book, ten usability factors and seven usability criteria have been developed as follows:

1. *Efficiency*: the capability of the software product to provide appropriate performance, relative to the number of resources used under stated conditions.
2. *Effectiveness*: the capability of the software product to enable users to achieve specified goals with accuracy and completeness in a specified context of use
3. *Productivity*: the capability of the software product to enable users to expend appropriate amounts of resources in relation to the effectiveness achieved in a specified context of use.
4. *Satisfaction*: the capability of the software product to satisfy users in a specified context of use.
5. *Learnability*: the ease with which users can master the required features for achieving their goals.
6. *Safety*: the capability of the software product to limit the risk of harm to people or other resources.
7. *Trustfulness*: the degree of faithfulness a software product offers to its users.
8. *Accessibility*: the capability of the software product to be used by permanently or temporarily disabled persons (i.e. vision, hearing, motor, cognitive, and language impairment).
9. *Universality*: the capability of the software product to accommodate a diversity of users (e.g. cultural aspects).
10. *Usefulness*: the degree to which a software product actually helps to solve users' practical problems.

Each factor is broken down into criteria as follows:

1. *Minimal action*: the capability of the application to help users achieve their tasks in a minimum number of steps.

2. *Minimal memory load*: whether users are required to keep a minimal amount of information in mind in order to achieve a specified task.
3. *Operability*: the amount of effort necessary to operate and control an application.
4. *Privacy*: whether users' personal information is appropriately protected.
5. *Security*: the capability of the application to protect information and data so that unauthorized persons or systems cannot read or modify them and authorized persons or systems are not denied access.
6. *Load time*: the time required for the application to load (i.e. how fast it responds to users).
7. *Resource safety*: whether resources, including people, are handled properly, without any hazard.

Consider this example to illustrate the applicability of the usability factors and their corresponding criteria using an MTM as shown in Table 7.1. It demonstrates nine usability factors and seven usability criteria. For example, the relation between the usability factor *efficiency* is assumed to correspond to the criteria *minimal action, operability, privacy, resource safety,* and *minimal memory load*.

7.1.2.1 User Authentication Use Cases

The following user authentication use cases have been developed to serve as the task scenarios:

■ Authenticate to an MTM and sub-systems
■ Transfer funds to an international bank account
■ Buy a concert ticket
■ Access an MTM with a mobile phone
■ Deposit a check using checking image
■ Send a silent alarm

The details for "Authenticate to an MTM and sub-systems" are presented in Table 7.2.

7.1.2.2 Demonstrating the Usable Security Symmetry Inspection Method using a Multifunction Teller Machine

The good old Automated Teller Machine (ATM), which is an electronic telecommunications device that enables customers of a financial institution to perform financial transactions, particularly cash withdrawal, without the need for a human cashier, clerk or bank teller. These ATMs cannot defeat what the authors designate

Table 7.1 Usability Factors and Usability Criteria Mapping for a Multifunction Teller Machine (MTM)

#	Task Scenario	Security Problem/Threat	Usability Criteria	Usability Factors								
				Efficiency	Satisfaction	Productivity	Learnability	Safety	Trustfulness	Accessibility	Universality	Usefulness
1	Authenticate to an MTM and sub-systems.	Storage of information: Replay attacks, eavesdropper, session hijacking, man-in-the-middle and verifier impersonation.	Minimal Action	●	●		●			●		
2	Transfer funds to an international bank account.	Access Control.	Operability	●	●				●	●		●
3	Buy a concert ticket.	Sensitive Information.	Privacy	●	●				●	●		●
			Minimal Action	●	●		●			●		

(Continued)

Table 7.1 (Continued) Usability Factors and Usability Criteria Mapping for a Multifunction Teller Machine (MTM)

#	Task Scenario	Security Problem/Threat	Usability Criteria	Usability Factors								
				Efficiency	Satisfaction	Productivity	Learnability	Safety	Trustfulness	Accessibility	Universality	Usefulness
4	Access an MTM with a mobile phone.	Credentials across several channels.	Security					●	●			●
5	Deposit a check using checking image.	Encryption.	Loading Time	●		●					●	●
6	Send a silent alarm.	User Physical safety.	Minimal Memory Load	●	●		●			●	●	●
			Resource Safety					●				

Table 7.2 Authentication Use Cases as Task Scenarios for a Multifunction Teller Machine (MTM)

Use case				Usability		Security	
#	Name	Scenario	Required Features	Problem	Scenario	Problem	Scenario
1	Authenticate to a Multifunction Teller Machine (MTM) (and its sub-systems).	Customer must log in using a multipurpose contactless (*) token-based authentication in combination with knowledge-based method (PIN) in order to have access to the MTM. The Multipurpose contactless card (MpCC) provides access to other services than those supplied by the MTM such as keeping medical records, physical access to buildings (when used with a reader), and making small purchases with an electronic purse.	**To log into the MTM:** • Insert your MpCC in the MTM's card slot; • Enter your PIN. **To authenticate to a medical institution logging onto a computer:** • Insert your McCC in the smart card reader; • Enter your PIN. **To authenticate to a facility (e.g. building) that has physical access control which authenticates individuals and permits access to physically secure areas:** • Bring your MpCC close to a card reader.	User Convenience (multipurpose vs. single purpose smart cards).	Deploying multiple applications on a single card (MpCC) not only saves time and money for organization and user alike, but it also simplifies the user's life. However, it raises the risk if the card is lost, stolen, or forgotten by users. Using a single purpose card is more secure, but it means users need to carry one card for each application, which is not so convenient.	Storage of Information	A MpCC, however, puts more sensitive information on the card (i.e. having all applications in one card is clearly less secure than a single card with a single application) and also requires more complex organizational coordination. The risk involved if the wrong person gets access to the card is much higher. Moreover, MpCCs open the door to attacks which exploit over-the-air communication channels in an unsolicited way

(Continued)

Table 7.2 (Continued) Authentication Use Cases as Task Scenarios for a Multifunction Teller Machine (MTM)

Use case			Usability		Security	
#	Name	Scenario	Required Features	Problem	Scenario	Problem	Scenario
1	(*) A contactless smart card includes an embedded smart card secure microcontroller, internal memory, and a small antenna; it-cates with a reader through a contactless radio frequency interface.	**To debit an amount from an electronic purse** (i.e. the card can be loaded with "electronic" value that can be decremented as purchases are made): **Pay your lunch with your multipurpose contactless smart card in the cafeteria.**				*Eavesdropping (a man-in-the-middle attack, where someone might cut out a few data blocks and replace them with other data to change, for example, transaction amounts) and interruption of operations*	
1						The card moves in the electromagnetic field; thus, the communi-cation between the reader and the card may be interrupted at any time without notice.	

Table 7.2 (Continued)   Authentication Use Cases as Task Scenarios for a Multifunction Teller Machine (MTM)

Use case				Usability		Security	
#	Name	Scenario	Required Features	Problem	Scenario	Problem	Scenario
2	Transfer funds to an international bank account.	Users must prove their identity again to protect access to particularly sensitive applications (i.e. authentication for high-value financial transactions) through a biometric-based authentication method: palm recognition.	When prompted, users place their hand on the palm scanner. Three-factor authentication (3FA) is the use of identity-confirming credentials from three separate categories of authentication factors – typically, the knowledge, possession, and inherence categories.	Heavy Workload	Users have to deal with both access control and strong authentication, that is, in this case, three-factor authentication such as *biometric authentication, KBA, and smart card. Biometric authentication (palm scanning):* User is not habituated to biometric identification.	Access Control	In a high-value financial transaction environment, procedures to control access to several areas of the card become predominantly important. The degree of security changes with the degree of sensitivity of the data related to the application. The issue of data security becomes more complex, for example, in a high-value financial transaction because the application requires a different level of security. Some applications may require no security; others may be sufficiently protected by a PIN;

(Continued)

Table 7.2 (Continued) Authentication Use Cases as Task Scenarios for a Multifunction Teller Machine (MTM)

Use case				Usability		Security	
#	Name	Scenario	Required Features	Problem	Scenario	Problem	Scenario
		Palm scanner embedded in the software on the MTM. Palm scanner placed outside and to the right of the MTM.			*KBA (Password/PIN):* This user has already been identified by a two-factor authentication method. However, this is a high-value financial risk transaction so user has to prove again his identity to the system. *Smart Card:* user has something *you have.*		others may require the use of biometrics to protect access to particularly sensitive applications, which is the case here. Thus, control of access to resources in a system is an important feature in order to provide integrity, confidentiality, and availability; it consists of two parts: verifying the claimed identity of the customer (authentication), and, thereafter, giving a properly

(Continued)

Table 7.2 (Continued)   Authentication Use Cases as Task Scenarios for a Multifunction Teller Machine (MTM)

Use case				Usability		Security	
#	Name	Scenario	Required Features	Problem	Scenario	Problem	Scenario
							authenticated customer access to the right resource (authorization) to which this particular one has restricted access (i.e. transfer funds to an international account). The biometric feature is another security layer to the current system. The palm recognition authentication system is build up around taking a three-dimensional view of the hand in order to determine the geometry and metrics of such variables as the length and thickness of fingers.

(Continued)

Table 7.2 (Continued) Authentication Use Cases as Task Scenarios for a Multifunction Teller Machine (MTM)

Use case			Usability		Security	
#	Name	Scenario	Required Features	Problem	Scenario	Problem	Scenario
							How does it work? First, an infrared scan is taken of the veins on the user's palm and data from the scan is stored in an integrated circuit chip embedded in the cash card or directly at the MTM biometric mechanism. When using the MTM, the users place their hand on the palm scanner, which verifies whether the pattern of veins matches that stored on the card or in the MTM. Hand geometry can only be used for verification.

(Continued)

Table 7.2 (Continued) Authentication Use Cases as Task Scenarios for a Multifunction Teller Machine (MTM)

Use case				Usability		Security	
#	Name	Scenario	Required Features	Problem	Scenario	Problem	Scenario
3	Buy a concert ticket.	After the user made a bank transaction, s/he decides to buy a concert ticket that s/he has just seen in the "Buy a Concert Ticket" section at the MTM. This user has already bought a concert ticket from an MTM once. At that time, she entered sensitive information such as her credit card number. However, now the MTM requests again the same user to enter her sensitive information.	• Tap on "Buy a Concert Ticket" section; • Select the number of seats; • Select credit card brand; • Tap in credit card number; • Tap in credit card's month and year of expiration.	Cumbersome user input requirements.	User has to enter sensitive information (i.e. the credit card number) each time she purchases a concert ticket. It does not provide convenience for the customer. The customer should be able to enter sensitive information only once at registration process.	Sensitive Information	Should the user enter sensitive information only at registration or each time a concert ticket is purchased? The first method makes a "one-click" feature" possible, meaning the user only has to select this option to order the ticket, providing convenience for her. However, the second method is better from a security perspective, but it requires more activity on the user's part.

(Continued)

Table 7.2 (Continued)   Authentication Use Cases as Task Scenarios for a Multifunction Teller Machine (MTM)

#	Name	Scenario	Required Features	Usability Problem	Usability Scenario	Security Problem	Security Scenario
4	Access an MTM offline with a mobile phone.	User accesses the MTM via mobile phone in order to make his monthly mortgage payment. The mobile phone is equipped with a special chip that enables to communicate with the MTM.	• Select "Access my MTM" from my mobile phone menu; • Enter 4-digit PIN (it is entered on the user's mobile phone keypad then transmitted to a central server and checked against file saved there); • Select "Make a Payment" from the MTM's menu; • Select the type of payment ("Mortgage"); • Tap the exact amount; • Select "Submit".	Overwhelms users with complexity when dealing with different communication channels.	Users have to manage complexity when dealing with different services offered through different types of communication channels such as MTM, Web, and wireless. Although it might be a convenient service when one does not have access physically to an MTM, it does place the burden on the user related to the coordination of the MTM and mobile phone. In addition, the user will still be required to authenticate to	Credentials across several channels	Using the same authentication credentials for both wireless and MTM channel can provide convenience for users. However, the PIN code is the only acceptable alternative for the wireless channel and is not secure enough (i.e. longer PINs (6 or 8-digit PINs) would be more secure than 4-digit PINs). Also, when the PIN is used for authentication over the mobile phone, the risk of eavesdropping this channel is an extra threat, especially since it cannot be encrypted.

(Continued)

Table 7.2 (Continued) Authentication Use Cases as Task Scenarios for a Multifunction Teller Machine (MTM)

| Use case | | | | Usability | | Security | |
#	Name	Scenario	Required Features	Problem	Scenario	Problem	Scenario
4					the system by entering a PIN. Unlike passwords, PINs have no meaning to the customer, and then it might be even harder to remember than a password (i.e. passwords can be created to be pronounce-able). PINs become harder to remember for users who have many different ones to keep track of.		

(Continued)

Table 7.2 (Continued)   Authentication Use Cases as Task Scenarios for a Multifunction Teller Machine (MTM)

Use case			Usability		Security	
#	Name	Scenario	Required Features	Problem	Scenario	Problem	Scenario
5	Deposit a check using checking image.	User has to deposit a cheque where the MTM accepts envelope-free check deposit.	• Select "Deposit a Check"; • Select "Scan my Check"; • Select if you want a paper or electronic receipt (i.e. it will be saved in your checking account).	Lower response time to the user.	Although it might be considered to reduce "perceivably" transaction speed (i.e. deposit a check without envelope), it might take more time upfront to process the data (i.e. capture the check's image) and the speed with which the MTM responds to the user. If the checking image seems slow, users might be less inclined to use it.	Strive for security using encryption.	The speed with which the MTM captures the check's image and responds to the customer is very important. However, in this case, the check's image must ideally be encrypted to provide secure communication and storage of information. While improving security, encryption takes time and makes the system slower, which can be very inconvenient for users. There is a trade-off between having a secure communication and having a short response time.

(Continued)

Table 7.2 (Continued) Authentication Use Cases as Task Scenarios for a Multifunction Teller Machine (MTM)

Use case				*Usability*		*Security*	
#	*Name*	*Scenario*	*Required Features*	*Problem*	*Scenario*	*Problem*	*Scenario*
6	Send a silent alarm from the MTM.	User has a considerable amount of money in his hands which is going to be deposited in his checking account. He starts the process of authentication (but didn't finalize it) when he notices a suspicious person approaching him. The intention of this person is obvious: to steal the user's money.	User enters an emergency PIN at the MTM alerting the bank's security system.	Difficulty remembering PIN or password especially under pressure.	People have even more difficulty remembering PINs and passwords, especially under pressure. In the case of money of a user being stolen, the threat is clear.	User security while using MTMs.	Adoption of an emergency PIN system at an MTM. User is able to send a silent alarm in response to a threat and get help from the bank. This can be the case, for example, of a user who has subscribed for this type of service with his bank.

as the next generation of ATMs, the Multifunction Teller Machine in several aspects, especially in the "transaction" aspect. Being able to perform up to 150 kinds of transactions ranging from straightforward cash withdrawals and deposits to fund transfer, to trading in stocks, to purchasing mutual funds, to cashing a check using check imaging, to something as ordinary as processing the payment of electricity bills, booking air-tickets, purchasing concert tickets, and making hotel reservations. An MTM is, in effect, the next generation of an ATM; it is a fully integrated cross-bank MTM network providing functionalities which are not straightforwardly associated to the management of one's own bank account, such as loading monetary value into pre-paid cards (e.g. mobile phones, tolls, service and shopping payments, etc.). Also, the MTMs can provide advanced authentication capabilities such as biometrics (e.g. palm recognition). For example, Bank of America is rolling out Teller Assist Machines (TAMs) that allow customers to video chat with bank tellers and receive exact change down to the penny, or video chat with a live teller, or get exact change, or pay credit card bills. On these new TAMs, for example, users can swipe their debit card, credit card, driver's license, or photo ID for authentication and then a live teller – physically based in Delaware or Florida – pops up on the screen to assist them.

To illustrate how the usable security symmetry inspection method can be applied in a real-world application, and how inspection method elements such as usability and security factors were selected and determined for the MTM example, the authors depict one of the use cases described in the previous section: "Transfer funds to an international bank account", a three-factor authentication. This symmetry inspection method has been applied to determine which type of authentication method should be used for this particular use case.

A user, Bob, needs to transfer $5,000 to an international account by dealing with both access control and strong authentication. He first authenticates himself to the MTM using a smart card and a PIN (the bank PIN policy states that a PIN must have 4 digits and 1 letter). In a high-value financial transaction environment, procedures to control access to several areas of the card become predominantly important; the degree of security changes with the degree of sensitivity of the data related to the application. The issue of data security becomes more complex in a high-value financial transaction because the application requires another layer of security to the current system: Biometrics. As this represents a high-value transaction, the MTM asks Bob to prove again his identity. So in addition to the bank card and PIN, Bob must employ a biometric authentication such as palm recognition – a three (multiple)-factor authentication.

7.1.2.3 The Usable Security Symmetry Inspection Method

The goal of making a system secure and usable will be successful only if it is a pre hoc consideration. This strengthens the argument made by other HCI-SEC researchers (Balfanz et al., 2004; Flechais et al., 2003; and Yee, 2004) that security and usability must be developed in unison from conception right through to development as an integral part of the system if they are ever to align perfectly. The authors agree with Yee (2006) that security and usability not only must be taken into consideration early and iteratively but also *together*. According to Yee (2006), integrated iterative design means iterative development processes based on repeated analysis, design, and evaluation cycles, rather than linear processes in which security or usability testing occurs at the end. Although many teams have adopted iterative processes, few seem to incorporate security and usability throughout. Not only is it important to examine these issues early and often, it is vital to design the UI and security measures together. Iterating offers the opportunity to see how security and usability decisions affect each other. Moreover, since usable security requires UI design requirements that are not similar to those of universal consumer software, it should also require usability evaluation methods that are appropriate to security. Standard usability evaluation methods could possibly treat security functions as if they were primary rather than secondary goals for the user, leading to flawed conclusions.

Symmetry also plays an important role in understanding the framework of this inspection method and how it has been built. Among a variety of definitions of symmetry, a generalized concept of the term has been adopted which is defined as follows: "Symmetry is a relationship of characteristic correspondence, equivalence, or identity among constituents of an entity or between different entities". The entities are defined as *security* and *usability*, and the mentioned relationship is the final outcome, which is the *usable security*. Another (notable) definition which relates to the definition of symmetry for the purposes of this book is from Weyl (1952), who states that: "Symmetry, as wide or as narrow as you may define its meaning, is one idea by which man through the ages has tried to comprehend and create order, beauty, and perfection." The word *order* is, in fact, a synonym of harmony. The utmost goal is that security and usability will no more be two separate entities but will work in harmony to produce secure and easy-to-use user authentication methods.

7.1.2.3.1 Definition

The *usable security symmetry* inspection method is a usable security inspection method which involves having a group of evaluators systematically examine a user interface and judge its compliance with security and usability principles. *Interface* is regarded, in this book, as both software components (e.g. user logs into a website) and hardware components (e.g. authentication token) toward which the interaction and information transit between software and/or hardware components, networks, and users.

This inspection method can be used to *guide a design decision* or to *assess a design* that has already been created. It integrates usable security earlier into the requirements and design phase, which helps security designers make more informed and therefore better decisions as well as influence the design in its early stage when, traditionally, the bulk of the feature design is done.

As the usable security symmetry inspection method provides very specific and practical review questions (not general ones), it is common to unfold issues and opportunities for feature improvement other than only those related to security and usability. This can be helpful when designing an authentication method, and it is in fact much appreciated given that we are talking about a computer security application.

According to Nielsen (1992), usability specialists were much better than those without usability expertise at finding usability problems by heuristic evaluation. Moreover, usability specialists with specific expertise (e.g. security) did much better than regular usability specialists without such expertise, especially with regard to certain usability problems that were unique to that kind of interface. Thus, this inspection method is developed as a usable security inspection method for system designers (acting also as evaluators) who have knowledge in computer security (especially user authentication) and a general knowledge of usability techniques and requirements.

7.1.2.3.2 Usable Security Protocol (USP) Sub-Methodology

The usable security symmetry inspection method is a sub-methodology within the USP methodology; therefore, the next sections continue on from the work described in that section. As mentioned, the USP methodology details step by step how the symmetry inspection method is created and generated, including goals, logistics, and the content behind each step. Also, it shows how the USP methodology generates, as the outcome, the design requirements inspection method tool for the design of user authentication methods: the usable security symmetry.

7.1.2.3.2.1 Inspection Method Category

The usable security symmetry is a checklist-based inspection method. A checklist is a valuable evaluation method when carefully developed and applied. A robust evaluation checklist clarifies the criteria that as a minimum should be considered when evaluating something in a particular area; aids the evaluator not to forget key criteria; and, finally, enhances the assessment's impartiality, reliability, and reproducibility. Another relevant benefit of employing checklists lies in the fact that they offer an organizational framework for quick recall of critical information and current best practices.

Such a checklist is useful in the authentication method lifecycle process such as planning an authentication method, monitoring and guiding its operation, and assessing its outcomes.

Moreover, checklists are useful for both formative and summative evaluations (Stufflebeam, 2000). Our symmetry inspection method makes use of the *formative evaluation*, which is a process of ongoing feedback on performance. The purposes are to specify aspects of performance that need improvement and provide corrective suggestions. *Summative evaluation*, in turn, is a process of specifying larger patterns and trends in performance and judging these review statements against criteria to get performance ratings.

A snapshot of our symmetry inspection short form checklist is shown in the following. Depending on the evaluation being carried out, the checklist can be quite long, so users should be able to collapse or expand each checklist item (e.g. 1.1, 1.2, 1.3. . . N), thereby facilitating data visualization as shown in Table 7.3. The long form checklist will show all rows expanded as can be seen in the next paragraphs.

The sub-methodology is described as follows:

- ■ Project lead activities:
  - – Identify and define the seven usability criteria that will be used to evaluate the authentication method.
    - • Minimal action: Capability of the application to help users achieve their tasks in a minimum number of steps (i.e. the length of transactions and procedures). It concerns perceptual and cognitive workload for individual inputs or outputs.
    - • Operability: Amount of effort necessary to operate and control an application.
    - • Privacy: Whether users' personal information is appropriately protected.
    - • Security: Capability of the application to protect information and data so that unauthorized persons or systems cannot read or modify them and authorized persons or systems are not denied access
    - • Load time: Time required for the application to load (i.e. how fast it responds to the user).
    - • Minimal memory load: Whether a user is required to keep a minimal amount of information in mind in order to achieve a specified task.
    - • Resource safety: Whether resources (including people) are handled properly, without any hazard.
  - – Identify one (or up to five) security designers and/or usability professionals to examine the system on an individual basis.
  - – Gather materials that facilitate the evaluators to become familiar with the purpose of the system and of its users (e.g. system specification, user tasks, personas, use case scenarios, etc.).
  - – Gather and analyze primary, secondary, and tertiary data available to build the inspection method.
- ■ Development activities:
  - – Develop the review questions in conjunction with the occurrences.
    - • Review questions: **Asking questions is a crucial component of finding information. The questioning method adopted for our inspection method is a combination of review and survey questions**

Table 7.3 Example of Data Visualization for the Inspection Method Checklist

Usable Security Symmetry Inspection Method

1. Usability Criterion: Minimal Action

The capability of the application to help users achieve their tasks in a minimum number of steps (i.e. the length of transactions and procedures). It concerns perceptual and cognitive workload for individual inputs or outputs.

#	Usability Review	Occurrence			Comments	Security Review	Occurrence			Comments
		Y	N	NA			Y	N	NA	
1.1	Can the user ...	▓			Users could do...	Is a single ID ...	▓			—
1.2	Is the workload ...			▓	—	Have strict ...	▓			Security policy ...
1.3	Is the authentic ...	▓			It seems pretty...	Is authentication ...	▓			—
1.4	Has the user to ...		▓		This can take...	Does the permis...	▓			—

which has been named by the authors as "review question". A review question presents users with a question or statement to which they answer using **a predefined set of scales (see occurrences section below). The** goal of the **review question is to** sum up and ask for agreement or otherwise.

- The review questions should be developed by taking into account the following quality guidelines:
  - Ensure each review question is completed properly with the ultimate goal of providing usable security for user authentication.
  - Make certain that the review question is pertinent to each specific usability criterion (e.g. review question for minimal action: *Has the requirement of data entry by users been limited when the data can be derived by the application or browser?*).
  - Include the cognitive process involved, implicitly or explicitly.
  - Identify particular error-prone areas (e.g. review question for minimal action: *Should PINs longer than six digits be avoided?*).
  - Make use of nomenclature that is well known in the domain (computer security and usability) for efficient communication.
  - Consider the expertise of the computer security and/or usability experts (i.e. target user).
  - It is important to note that the review questions are not an exhaustive list of all usability and security issues but rather the common ones. Also, since technology constantly evolves, so should the review questions.
- **Occurrences** are represented by the following attributes: **Y** (Yes), **N** (No), and **NA** (Not Applicable).

  Y represents that the occurrence complies with the review question; **N** represents that the occurrence does not comply with the review question; and finally, **NA** represents that the occurrence is not applicable with the review question. The default value for the occurrence field is empty (none).
  - Each attribute has an associated value. If the attribute is **Y**, then its value is a green color, when applicable. If **N**, then a red color, and finally, if **NA**, then a gray color. For easier data visualization, it is recommended to use colors then symbols or numbers. For visually impaired people, some alternatives are gray scales, checkmarks, numbers, etc.).

- **The Comments** attribute represents any comments that evaluators consider relevant to the review questions. If there are no comments, a dash "—" (or any other sign meaning "no data") should be entered in the respective value field since leaving it blank may mislead evaluators into thinking that data is missing.
- Evaluate the system:

- Review the materials provided to become familiar with the system design. If possible, carry out the user actions that will be taken to perform the user tasks.
- Identify and list any areas of the system that might be opposed to the usability and/or security principles. List all of the concerns that you note in the "Comments" field.

■ Show the outcome:
- The main outcome of the inspection method is a list of usability and security issues in the interface with references to those usability criteria and security aspects previously described in the Usability Severity Ratings and Recommendations for MTM and Security Severity Ratings and Recommendations for MTM sections.

■ Analyze the results:
- Review each of the concerns that have been written down in the evaluated checklist.
- Assess and judge each concern for its compliance with your defined criterion.
- Allocate a severity level for each grouped concern based on the impact to the end user (See the section about Severity Ratings).
- Establish recommendations to fix the problem. Ensure that each recommendation relates to the criterion.
- Finally, if needed, use a holistic approach to review all recommendations and verify that there are no major (or none at all) conflicts among them. If there are any, make an assessment and implement only those recommendations that address high-priority functional system requirements.

The inspection method is shown in its integrity in Table 7.4.

Table 7.4 Usable Security Symmetry Inspection Method

#	Usability Review	Occurrence			Comments	Security Review	Occurrence			Comments
		Y	N	NA			Y	N	NA	
1. Usability Criterion: **Minimal Action**										
The capability of the application to help users achieve their tasks in a minimum number of steps (i.e. the length of transactions and procedures). It concerns perceptual and cognitive workload for individual inputs or outputs.										
1.1	Does the system employ multiple-technology[1] contactless[2] cards for physical and logical[3] access applications?	▓			Current system adopts a contact card. All-in-one contactless card would be nice to have. Very convenient for users struggling with several plastic cards in their wallets.	For stronger security, has a third-factor authentication been introduced (e.g. biometrics)?	▓			Hand geometry recognition is used for authentication in conjunction with the smart card plus PIN. *Physical Access:* Plastic cards will be obsolete since contactless technology is growing with a variety of form factors (e.g. take payments on your smartphone/tablet using the Square reader with a Square Register app, or open doors, access subway, etc.) and acting as another form of authentication.
1.1			▓					▓		*Logical Access:* Currently, contact technology provides a convenient, quick, and cost-effective way to transfer considerable amounts of data between a card and a reader/host system. Also, it performs complex cryptographic operations for authentication applications. Therefore, contact smart cards are a prominent solution for network security implementations.

(Continued)

Table 7.4 (Continued)   Usable Security Symmetry Inspection Method

#	Usability Review	Occurrence			Comments	Security Review	Occurrence			Comments
		Y	N	NA			Y	N	NA	
1.2	Is the cognitive workload low and simple (e.g. input workload kept to a minimum)?				Plenty of user interaction when dealing with the different MTM functionalities. Perhaps defining a simpler feature set can mitigate this issue.	Have strict password policies been established (i.e. one-factor authentication (1FA) like password, a strong password policy must be set up)?				Users might have to authenticate through several successive stages to access their resources. Access to some functionalities in the MTM may require various levels of proofs, depending on the risk level of the transaction, which in turn requires higher levels of security such as first, second, third and so on levels of authentication (e.g. a combination of password, smart card, and palm recognition).
1.3	Is the authentication process simple for users (i.e. verify and authenticate the identity of users)?				If there's only first level authentication (smart card and password); if users need more privilege or access critical services it may need a second level authentication (e.g. hand recognition)	Is authentication of principals to resources required?				—

(Continued)

Table 7.4 (Continued)   Usable Security Symmetry Inspection Method

#	Usability Review	Occurrence			Comments	Security Review	Occurrence			Comments
		Y	N	NA			Y	N	NA	
1.4	Has the user to re-authenticate when effectuating transactions other than financial (e.g. buying a concert ticket with a credit card).	▨			Re-authentication is cumbersome to users.	In a re-authentication scenario, are the different authentication methods integrated?		▨		System review and implementation. Re-authentication forces a user to perform an additional login to ensure that a user who is accessing a protected resource is the same person who initially authenticated at the start of the session. Re-authentication provides additional protection for sensitive resources in the secure domain.
1.5	Has the user to re-authenticate when accessing information from other channels?	▨			This is really bothersome and bad user experience to users.	Are security requirements adequate and balanced with users' tasks?[4]	▨			This depends on integrating a role-based authentication which includes access control mechanism, which grants and revokes privileges based on predefined rules.
1.5.1	From web-based application?	▨			Users have already logged into the MTM with the conventional smart card plus PIN. Using another authentication method (e.g. OTP) really becomes a heavy workload to users.	Is strong authentication used with at least two-factor authentication (2FA)?	▨			Smart card plus PIN: Smart cards require a PIN so they add a second layer of security (smart card/PIN in place of a password) that an impersonator would have to obtain to log onto a system.

(Continued)

Table 7.4 (Continued) Usable Security Symmetry Inspection Method

#	Usability Review	Occurrence			Comments	Security Review	Occurrence			Comments
		Y	N	NA			Y	N	NA	
1.5.2	From Electronic Purse (EP)? A smart card becomes an electronic purse when money is credited to thecard via an electronic reader.				It seems a little tiresome to users carrying another plastic card, but EP is less bulky than cash and is very convenient to users.	Is strong authentication used with at least 2FA?				EP is more secure than cash since they can be "closed" by a single keystroke and reopened by entering a four-digit PIN. However, the advantage of the EP is really in its comparison with cash: once users have obtained their cash from an MTM, anyone can spend it.
1.6	Is the user interface content of the application (i.e. information architecture, widgets, dialog windows, copy, etc.) displayed succinctly (i.e. shorter reading times, smaller errors, etc.)?				—	Is required content associated with sensitive information kept to a minimum?				Basically, username and email address are required.
1.7	Is the password usable (i.e. meaningful and concatenate/ interspersed words with characters)?				This depends on a predefined password policy. Users can choose a meaningful and concatenate password as long as it respects the standard password policy.	Do passwords follow strict security policies?				Password policy requires a strong password which usually must contain at least eight characters, one uppercase alphabet (A–Z), one lowercase alphabet (a–z), one Arabic numeral (0–9), one non-alphanumeric character excluding @~–#&^%.

(Continued)

Table 7.4 (Continued) Usable Security Symmetry Inspection Method

#	Usability Review	Occurrence Y	Occurrence N	Occurrence NA	Comments	Security Review	Occurrence Y	Occurrence N	Occurrence NA	Comments
1.8	Is a particular cue given to users for remembering passwords (e.g., placing a post label next to the password field such as "Your password is your phone number")?	▨			A cue can be defined as: 1) A reminder or prompting, or 2) A hint or suggestion. A cue heard by a potential intruder, therefore, could help the intruder as much as the legitimate user, especially if the cue is easily understood.	Are text field post labels that provide tips for remembering passwords and PINs avoided?		▨		One option to mitigate the issue of forgetting a password is to display a "Forgot My Password" link on the UI so that users can reset it without being forced to remember a password.
1.9	Is Single Sign-On (SSO) login system used?		▨		To remember a single password for a variety of applications available in a system, SSO is appropriate. As mentioned, memorability is further improved through the use of mnemonics to aid their recall (Miller, 1967). Mnemonics are memory devices that help learners recall larger pieces of information, especially in the form of lists like characteristics, steps, stages, etc. Interoperability is very difficult for now.	Are strong passwords strictly required when using SSO?	▨			Use also password mnemonics: using mnemonics to create and remember super-strong passwords. It's a method which uses, for example, lyrics such as "Hey Jude, don't make it bad" or "HJdmib".

(Continued)

Table 7.4 (Continued)   Usable Security Symmetry Inspection Method

#	Usability Review	Occurrence			Comments	Security Review	Occurrence			Comments
		Y	N	NA			Y	N	NA	
1.10	Have data entry user requirements been as restricted as possible when data is derived by the application or browser?				Only for second level authentication, when different types of authentication methods will be presented to users since they are accessing "external" online web applications. Most common is username/password: usually, it saves credentials in a cookie so the user doesn't need to login every time.	Is the use of default "cookies"5 avoided when using 1-factor authentication (1FA) (e.g. username/ password)?				Storing credentials in a default cookie has the disadvantage that everyone who has access and knowledge about the MTM can very easily see the user's password by looking at the stored cookies.
1.11	Does the system speed *data entry* by setting *default values?*				Although default values are different from using a cookie for username and password, it is somewhat related to 1.10	Are default values avoided, especially for sensitive information (e.g. username and password)?				Although default values are different from using a cookie for username and password, it is somewhat related to 1.10

(Continued)

Table 7.4 (Continued)   Usable Security Symmetry Inspection Method

#	Usability Review	Occurrence			Comments	Security Review	Occurrence			Comments
		Y	N	NA			Y	N	NA	
1.12	Does the system accommodate both experienced and novice users (e.g. shortcuts are available to experienced users)?				After entering PIN at the MTM, expert users generally tap directly on the "Fast Cash" (shortcut) button instead of "Enter" so that they authenticate and at the same time they go directly to the fast cash function. Novice users usually tap on "Enter" first.	Are succinct error messages displayed for novice users, avoiding serious errors?				Succinct error messages are displayed for both novice and expert users.
1.13	Are PINs longer than six digits avoided?				Four-digit PIN is used.	Does the system use weak PIN change protocols?				*PIN Attacks:* Fraudsters are targeting the automated telephone banking or Voice Response Unit (VRU) systems of financial institutions to change or obtain PIN information. After obtaining a valid PIN, fraudsters can then make unauthorized withdrawals at ATMs. *Weak PIN Change Protocols:* Fraudsters utilize automated Voice over Internet Protocol (VoIP) services or other automated broadband Internet phone services to dial financial institution VRUs. With stolen account numbers and other personal information (i.e. social security number, mother's maiden name), fraudsters are able to bypass security questions or other

(Continued)

Table 7.4 (Continued) Usable Security Symmetry Inspection Method

#	Usability Review	Occurrence			Comments	Security Review	Occurrence			Comments
		Y	N	NA			Y	N	NA	
										verification processes. For caller ID validation, "caller ID spoofing" can be used to masquerade or change the caller ID display that is transmitted with the call to appear authentic. Fraudsters can then exploit vulnerable financial institutions that have an automated PIN change option and select a new PIN to fraudulently use at ATMs. Users circumvent PIN forgetfulness through insecure behaviors (e.g. writing down their PINs, make them all the same, or disclose them to friends and family).
1.14	Are PINs employed mostly for frequently used systems (i.e. infrequently used ones are the ones most forgotten)?				Four-digit PIN is used.	Are smart cards used in conjunction with PIN to provide strong authentica-tion?				—

(Continued)

Table 7.4 (Continued) Usable Security Symmetry Inspection Method

#	Usability Review	Occurrence			Comments	Security Review	Occurrence			Comments
		Y	N	NA			Y	N	NA	
1.15	Does the smart card provide interoperability across services?	■			Users sometimes get lost with so many options and also different levels of authentication and authentication methods.	Have users the option to choose a single PIN or separate PINs for each application?		■		PIN is required only once when users first login at the MTM.
1.16	If users must switch between different systems, are dual-interface[6] chip smart cards[7] used?	■			Users switch to web applications not from the card.	Can card issuers record and update appropriate privileges from a single central location?	■			—
1.17	Does the system provide users with the ability to be authenticated by group/role (e.g. a group of users (two or more) can authenticate with the same credentials)?			■	System review and implementation to evaluate the feasibility of this functionality in an MTM.	Is the Role-Based Access Control (RBAC)[8] used to manage users who have signed up for this service?	■	■		RBAC diverges from Access Control Lists (ACLs) used in conventional discretionary access control systems in that it assigns permissions to certain groups to specific operations.

(Continued)

Table 7.4 (Continued)   Usable Security Symmetry Inspection Method

#	Usability Review	Occurrence			Comments	Security Review	Occurrence			Comments
		Y	N	NA			Y	N	NA	
2. Usability Criterion: **Operability**										
Amount of effort necessary to operate and control an application.										
2.1	Can (some) systems' mechanisms be customized[9] by users to operate in certain ways (e.g. select a preferred authentication method)?				Supporting the ability for users to customize their experience (i.e. product customization) is adding value to the product or service. Some organizations offer different ways of authenticating to a system, but this is taking place more specifically in corporate environments (e.g. usually, authentication options are given to system administrators). End users (i.e. consumers) cannot select their preferred authentication mechanisms so far.	Does the system offer different combinations of authentication methods to users that include strong authentication (e.g. OTP plus biometrics, or contactless smart card plus password, etc.)?				If organizations could offer different authentication methods options to users, this would provide users with more convenience and also, in some way, more commitment to security. This means that giving a bit more user control would probably increase their commitment and understanding of what security represents to the organization.

(Continued)

Table 7.4 (Continued) Usable Security Symmetry Inspection Method

#	Usability Review	Occurrence			Comments	Security Review	Occurrence			Comments
		Y	N	NA			Y	N	NA	
2.2	Does the MTM system offer user personalization?		N		System review and implementation. An example would be an MTM that displays a list of users' last three transactions to make reordering the same services easy.	Does the system personalize only non-sensitive information and non-security-related content items?			NA	—
2.3	If a variety of authentication methods is required, are they the simplest ones from the user's perspective?	Y			—	Is third-factor authentication required to protect access to sensitive applications (e.g. biometrics)?	Y			When making an international funds transfer above $5,000 at the MTM, users make use of a smart card and password and palm recognition.

(Continued)

Table 7.4 (Continued)   Usable Security Symmetry Inspection Method

#	Usability Review	Occurrence			Comments	Security Review	Occurrence			Comments
		Y	N	NA			Y	N	NA	
2.4	Is the biometric method usable and easy to use?				Palm recognition.	Has the chosen biometric method the lowest false reject rate (FRR) (i.e. a legitimate user is rejected by the acquisition device)?				Users do like palm and fingerprint recognition biometric authentication methods because they are usually easy to use and not intrusive. From the organizations' perspective, they are low cost and generally reliable, with only a 0.1% FRR, which is one of the lowest in the biometrics industry.
2.5	Does the user's task have a light workload?				Relatively simple for the corporate and consumer average user. As mentioned, the system uses smart card plus PIN (first level authentication) then second level (username/password, hand recognition, or OTP).	Is the user authentication flow simple in order to avoid dangerous behaviors (e.g. sensitive data leakage)?				Generally speaking, smart card plus PIN takes two steps. For username/password, according to the GOMS analysis previously developed, there are precisely 19 steps (24.40 seconds comprising cognitive and motor processes). However, the core process takes three steps, i.e. the user inputs username, then password and taps on "Submit" button. For OTP, according to the GOMS analysis, precisely 21 steps (28.13 seconds comprising cognitive and motor processes).

(Continued)

Table 7.4 (Continued) Usable Security Symmetry Inspection Method

#	Usability Review	Occurrence			Comments	Security Review	Occurrence			Comments
		Y	N	NA			Y	N	NA	
2.6	Are the transactions' sequences smooth when varying the different levels of access?				—	When varying levels of access, is authentication or user validation required?				—

(Continued)

Table 7.4 (Continued) Usable Security Symmetry Inspection Method

#	Usability Review	Occurrence			Comments	Security Review	Occurrence			Comments
		Y	N	NA			Y	N	NA	
2.7	Are there varying levels of logical access (i.e. access to local software at MTM and external web applications)?				It depends on which services users are accessing. For example, buying a concert ticket requires going online whereas using a check image feature is a local feature which does not need another level of authentication.	Has the system considered a high-security computing environment (i.e. this is required for public key infrastructure (PKI) and/or biometrics services)?				As people are more concerned with privacy risks in biometrics systems, and MTMs do make use of network and distributed systems, the use of encryption is critical. The security of the biometric system relies on the integrity and authenticity of the biometric information, which can be accomplished using PKI once users have been enrolled. Users have access to the ATM software, which can be, for example, a Windows-based multi-vendor ATM software solution. It is an open standards platform that includes a built-in ATM application with standard transactions and functionality that can be deployed "as is" or extended to incorporate advanced functionality. Other OSs being considered nowadays are Linux and Android.

(Continued)

Table 7.4 (Continued) Usable Security Symmetry Inspection Method

#	Usability Review	Occurrence			Comments	Security Review	Occurrence			Comments
		Y	N	NA			Y	N	NA	
2.8	For biometrics, are local and remote accesses available to user authentication (i.e. if online systems are down during an outage or other emergency)?				—	Is the local access set up to default (i.e. no need to be transmitted over the network)?				System review and implementation.
2.9	Is enrollment data maintained centrally or locally (i.e. user convenience – a key issue in selecting card issuance procedures)?				Users should enrol once and this be served as the basis of future authentication across a range of systems.	Is a smart card used to store user's biometric data, including user profile and enrollment?				System review and implementation. Although users may see the benefit of leveraging an existing enrollment, they may be suspicious of potential data misuse.
2.10	Is biometric authentication used?				Biometrics provides user convenience.	For a transaction with a high risk of a hacker attack, is biometrics used?				Man-In-The-Middle (MITM) attacks using biometrics are more difficult to go through than the other authentication methods.

(Continued)

Table 7.4 (Continued) Usable Security Symmetry Inspection Method

#	Usability Review	Occurrence Y	Occurrence N	Occurrence NA	Comments	Security Review	Occurrence Y	Occurrence N	Occurrence NA	Comments
2.11	If biometrics adopted, is hand geometry authentication used?	▦			Hand palm recognition is easy to use and non-intrusive.	For stronger security, is hand geometry authentication method used?	▦			Hand geometry authentication is more accurate and secure than fingerprint biometric method10. More secure if used in conjunction with other technologies like KBA (e.g. password and PIN).
2.12	If hand geometry authentication is used, are the false reject rate (FRR) and false acceptance rate (FAR) balanced to provide more convenience to users?		▩		System review and implementation.	Is the hand geometry FRR about 1% and the FAR about 0.1%/11?		▩		System review and implementation. Currently, the system presents an FRR of 5% and a FAR of 3%.
2.13	Apart from the common smart card and PIN used in the MTM, is there a second level authentication method rather than biometrics in order to provide availability to users?	▦			One-time password (OTP) authentication method. Availability here refers to a number of potential privacy, personal, religious, cultural, and legal issues associated with the use of a biometric, and also people with disabilities.	Is the OTP authentication method a hardware token rather than software token when biometrics is not available?	▦			Hardware OTP token. An authentication factor is a piece of information used to authenticate or verify a person's identity. Among the four authentication factors that might be employed in combination to increase the level of security in the claimed identity of a user is: something you have, which is the hardware OTP token.

(Continued)

Table 7.4 (Continued) Usable Security Symmetry Inspection Method

#	Usability Review	Occurrence			Comments	Security Review	Occurrence			Comments
		Y	N	NA			Y	N	NA	
2.14	If users forget their PINs, can they reset them via a web-based portal[12] accessed on the MTM rather than in an issuance station?		▓		Users need to visit a bank branch *in person* or wait for at least ten business days to receive their PIN by mail. Resetting their PIN via the Web would ensure customer convenience and satisfaction.	If users forget their PINs, can they reset them via a web-based portal and their mobile phone?		▓		Not possible, as spelled out in the bank's security policy. Combination of the Web and mobile phone would actually increase security due to the two-factor authentication aspect (i.e. something you know (password) and something you have (phone)).
2.15	If biometrics is adopted, are mechanisms and software used the most recognized in the industry to facilitate user convenience?	▓			—	Does the biometric solution follow applicable standards (i.e. many biometric solutions use their own proprietary algorithms and processes)?	▓			—

(Continued)

Table 7.4 (Continued) Usable Security Symmetry Inspection Method

#	Usability Review	Occurrence			Comments	Security Review	Occurrence			Comments
		Y	N	NA			Y	N	NA	
2.16	Is the mean time between failures (MTBF) the lowest rate possible, improving user experience?[13]				System review and implementation.	Is the MTM error rate of *simple processing* errors[14] of one error per 100,000 transactions?				System review and implementation.
2.17	Is the mean time to repair (MTTR)[15] lower than an hour, improving user experience and availability of services?				System review and implementation.	Is the MTM error rate of *simple processing* errors[16] of one error per 100,000 transactions?				System review and implementation.
2.18	Is the total transaction time (TTT) to handle a biometric method equal to minimum 4 and maximum 10 seconds for a single user?				Currently, hand geometry takes 15–20 seconds.	Is the biometric template maintained in a smart card[17] providing highly secure and portable authentication of the cardholder's identity?				Biometric template is kept in a central repository, mirrored in distributed databases.

(Continued)

Table 7.4 (Continued)   Usable Security Symmetry Inspection Method

#	Usability Review	Occurrence			Comments	Security Review	Occurrence			Comments
		Y	N	NA			Y	N	NA	
2.19	Can users open accounts and apply for loans and credit cards using video teller technology?				System review and implementation. This functionality would be very convenient for users. An important reason for the MTM's continued importance in the banking infrastructure is that the technology can offer much greater functionality than before. Not only do users see the bill and check payments, but video teller technology means consumers can now carry out much richer transactions at the MTM that typically only previously occurred in the branch.	Is the video conferencing channel encrypted?				System review and implementation.
2.20	Can the MTM dispense $1, $5, $20, and $100 bills, instead of only $20 bills like traditional ATMs?				System review and implementation. This functionality would be very convenient for users.	Is the use of $100 bills avoided?				System review and implementation. Generally speaking, very high quality counterfeit $100 bills are some of the most widely distributed counterfeit American dollar bills.

(Continued)

Table 7.4 (Continued)   Usable Security Symmetry Inspection Method

#	Usability Review	Occurrence			Comments	Security Review	Occurrence			Comments
		Y	N	NA			Y	N	NA	
2.21	Is user personalization available on the MTM (i.e. personalized screens based on customer preferences)?				System review and implementation. Users can personalize the UI of an MTM to control the order of the menu options that it is displayed. For example, if the most frequently withdrawn amount is $40, this would be the first option listed, followed by the less frequently chosen amounts. Similarly, if balances are requested more frequently than statements, the balance option would be displayed first. Placing the most frequently used options near the top of the menu will reduce search times and make option selection less error prone. Also, personalize the "favorites" that appear highlighted on the screen based on each customer's MTM usage.	Can users personalize only non-sensitive information and non-sensitive security information?				Sensitive information examples are checking account number, savings account number, credit card number, etc. Sensitive security information might be systems security information, security directives, etc.

(Continued)

Table 7.4 (Continued)   Usable Security Symmetry Inspection Method

#	Usability Review	Occurrence			Comments	Security Review	Occurrence			Comments
		Y	N	NA			Y	N	NA	
2.22	Can users withdraw cardless cash using a mobile app?				Mobile and online is surging, but a recent Federal Reserve report shows the ATM is still, after the branch, the most commonly used method among consumers to engage with bank services. Consumers want more, and in order for the MTM to remain at the heart of retail banking, it is important for deployers to think about what they can offer their users. People are demanding more choice and functionality. Also, using a mobile app is convenient for users because they will be authenticating to the system by using a device and authentication method that they are familiar with.	If a mobile app is used, does the system require at least a two-factor authentication?				The most common way of interacting with a financial institution remains in-person at a branch, with 87 percent of consumers who have a bank account reporting that they had visited a branch and spoken with a teller in the 12 months prior to the survey. However, the second-most common means of access in the previous 12 months was using an ATM at 75 percent18, followed by online banking at 74 percent.

(Continued)

Table 7.4 (Continued) Usable Security Symmetry Inspection Method

#	Usability Review	Occurrence			Comments	Security Review	Occurrence			Comments
		Y	N	NA			Y	N	NA	
3. Usability Criterion: **Privacy**										
Whether users' personal information is appropriately protected.										
3.1	If a smart card is used, who owns the personal data stored on it?				The MTM's owner, which in this case is a bank. It might be a loyalty program provider including healthcare, entertainment, and transportation.	Is the user the "owner" of personal information (user is responsible for keeping personal information on the card up-to-date)?				System review and implementation.
3.2	Does the system owner provide easy access to the privacy policy?				System review and implementation.	Has information about privacy protection been included in the privacy policy (card acceptance agreement)?				—

(Continued)

Table 7.4 (Continued)   Usable Security Symmetry Inspection Method

#	Usability Review	Occurrence			Comments	Security Review	Occurrence			Comments
		Y	N	NA			Y	N	NA	
3.3	Does the system provide the user with the right of access to the information and a process for correcting errors?				—	Is the system's owner responsible for the security and accuracy of information?				—
3.4	Are users in control of their private information?				System review and implementation.	Is the system application complying with State or Federal laws (e.g. Regulation E, State Privacy Acts, etc. in the U.S.)?				However, the owner needs to do an update regarding current or potential laws.
3.5	If the card is lost, can the user go to a single location to repopulate the card?				—	Is there a client registry that provides pointers to all application owner databases for all applications active on the card[19]?				System review and implementation.

(Continued)

Table 7.4 (Continued) Usable Security Symmetry Inspection Method

#	Usability Review	Occurrence			Comments	Security Review	Occurrence			Comments
		Y	N	NA			Y	N	NA	
3.6	Does the system owner provide integration of multiple databases[20]?	▓			System review and implementation.	Does the system owner create "shadow" files of all transactions and route these at least daily to the application owner's remote database?		▓		System review and implementation.
3.7	Is the cardholder privacy maintained?				It refers to the security of the information of the credential ID. PKI is used (encryption)	Is "Match-on-card" technique used[21]?		▓		System review and implementation. This technique is portable.
3.8	Are users assured about transactions made related to loss of privacy and security for payments?	▓			These transactions are made through the web interface on the MTM. However, the system needs improvements in this area (review and updates).	Does the system provide private and secure payments for Internet purchases?	▓			—

(Continued)

Table 7.4 (Continued)   Usable Security Symmetry Inspection Method

#	Usability Review	Occurrence			Comments	Security Review	Occurrence			Comments
		Y	N	NA			Y	N	NA	
3.9	Do security protocols adapt their encryption policies based on the content of the data being encrypted (e.g. video encryption algorithms[22])?		▨		System review and implementation.	Are encrypted tunnels[23] or virtual private networks supported to provide confidential access over insecure wireless links?	▨			—
3.10	Are users insecure and reluctant about their sensitive information (i.e. system has to be aware of the impact of application/network on remote end users' experiences and uncover and resolve problems before rollout)?	▨			Especially because it is an MTM, no matter how much secure it is. However, this is changing given that MTMs are now able to provide basically the same user experience and basic functionalities related to sensitive information as a computer.	Does the system provide data corruption[24], detection and prevention, and backups for databases)?		▨		System review and implementation. For example, tamper-evidence for digital signature generation25 or transaction logging, auditing remotely for MTM journal of transactions.

(Continued)

Table 7.4 (Continued) Usable Security Symmetry Inspection Method

#	Usability Review	Occurrence			Comments	Security Review	Occurrence			Comments
		Y	N	NA			Y	N	NA	
4. Usability Criterion: **Security**										
The capability of the application to protect information and data so that unauthorized persons or systems cannot read or modify them and authorized persons or systems are not denied access										
4.1	If logical access is provided and users, for example, forget their password, is there any mechanism to avoid the system's blockage after three failed attempts to log on?[26]	▨			System review and implementation.	Is logical access provided for local and remote networks to prevent unauthorized programs/users from accessing confidential data?	▨			For example, private keys or confidential information in databases.
4.2	When using different communication channels, is PIN authentication used?		▨		When accessing MTM, Web, and mobile, PINs are easier to remember than passwords.	Is a 6-digit PIN used?				System review and implementation. PINs have a lower level of security since the number of possible combinations is lower[27].
4.3	When using KBA over the telephone, is PIN used instead of a strong password?	▨			System review and implementation. PIN is easier to remember than password.	Are PINs avoided to be used as the only authentication mechanism?	▨			—

(Continued)

Table 7.4 (Continued) Usable Security Symmetry Inspection Method

#	Usability Review	Occurrence			Comments	Security Review	Occurrence			Comments
		Y	N	NA			Y	N	NA	
4.4	Can the customer make use of multifunction smart cards which provide substantial convenience for them?	▨			—	Are financial and security applications placed in separate card platforms?		▨		System review and implementation. The combination of financial and security applications raises potential security risks and interoperability issues that must be addressed in a multi-application environment.
4.5	For logical access application, is biometrics provided enhancing usability?		▨		System review and implementation. On-card biometric match and on-card key generation.	For logical access application, is contact technology provided for network security implementations?	▨			This is related to contact cards. It is a convenient and cost-effective way to transfer significant amounts of data between a card and a reader and host system, and to perform complex cryptographic operations for authentication applications. Furthermore, contact chips have microcontrollers while contactless chips may or may not. For these reasons, contact smart cards have been an outstanding solution for network security implementations.

(Continued)

Table 7.4 (Continued)   Usable Security Symmetry Inspection Method

#	Usability Review	Occurrence			Comments	Security Review	Occurrence			Comments
		Y	N	NA			Y	N	NA	
4.6	Does the user trust the MTM's system owner?				See also #3.10	Does the cryptography employ algorithms approved under Federal Information Processing Standard 140–2?				System review and implementation.
4.7	Have users more than a few alternatives to authenticate to the system, improving availability and convenience of the system?				—	Does the system use a higher-end card to support digital signatures and/or biometric capabilities?				With enough memory and/or a crypto processor.
4.8	Do users have access to multiple authentication technologies such as PKI and biometrics?				—	Is PKI used?				This is secure digital certificates for logical access control, remote access, and encryption.

(Continued)

Table 7.4 (Continued) Usable Security Symmetry Inspection Method

#	Usability Review	Occurrence Y	Occurrence N	Occurrence NA	Comments	Security Review	Occurrence Y	Occurrence N	Occurrence NA	Comments
4.9	For the transmission of credentials over the phone, is the traffic passing over a telephone network encrypted28?	▓			This would increase user satisfaction and trust in the system.	Is an additional authentication method used in conjunction with PIN over the phone?			▓	System review and implementation. This would provide, therefore, a strong authentication

5. Usability Criterion: **Load Time**

The time required for the application to load (i.e. how fast it responds to the user).

#	Usability Review	Occurrence Y	Occurrence N	Occurrence NA	Comments	Security Review	Occurrence Y	Occurrence N	Occurrence NA	Comments
5.1	Is the computation imperceptible to users?	▓			For example, the Secure Socket Layer (SSL) protocol requires processing time for encryption and authentication. This cost is easily masked by the latency of loading web pages.	Is PKI used when a high percentage of users perform monetary transactions?	▓			—
5.2	When the computation takes longer, is a progress indicator used; if it is not possible to mask it?		▓		System review and implementation. Progress indicator: The full length of an operation can be determined and we can tell users how much of the process has been completed.	To reduce the workload of a security protocol, are lightweight cryptographic algorithms for various security functions used?	▓			For example, Secure Socket Layer (SSL) protocol

(Continued)

Table 7.4 (Continued)   Usable Security Symmetry Inspection Method

#	Usability Review	Occurrence Y	N	NA	Comments	Security Review	Occurrence Y	N	NA	Comments
5.3	Is the speed of transaction time-consuming?		▓		But there is some room for improvements, so perform system review and implementation.	Is PKI used when a high percentage of users transmit and/or receive data across open networks?	▓			—
5.4	Does the system use Java Card technology[29] in order to have the response time reduced?		▓		The time to read data from and write data to the chip has been reduced substantially.	Is the adopted smart card a high-performance one to reduce latency when executing the encryption algorithm?		▓		System review and implementation.
5.5	If imaging technology is used, is the processing time required to scan a live image (e.g. a check) less than 1 second?	▓			—	Has an adequate period of time been allotted for providing encryption capabilities?	▓			—

(Continued)

Table 7.4 (Continued) Usable Security Symmetry Inspection Method

#	Usability Review	Occurrence			Comments	Security Review	Occurrence			Comments
		Y	N	NA			Y	N	NA	
5.6	Does the live image require a small memory allocation which will affect the time for retrieval of the image by the user?				—	Does the system provide dynamic memory allocation?				The dynamic memory allocation feature can maximize the concurrent execution of memory allocation routines and also optimize the allocation of both large and small blocks of memory.
5.7	Are the UI screens attractive and easy on the eye?				There are some areas for improvement.	Does the system make use of a Windows-based platform to get the best character recognition software?				Windows-based platform: Software package optimized for text acquisition with handheld and flatbed scanners.
5.8	When using check imaging technology, if the scanner rejects the inserted check because the handwriting is illegible, can users deposit it by the conventional way without fees?				It sends check images or information electronically to the system's owner. The conventional deposit is available at the MTM as well.	Has the check imaging process been integrated to the system owner's IT infrastructure?				System review and implementation.

(Continued)

Table 7.4 (Continued)   Usable Security Symmetry Inspection Method

#	Usability Review	Occurrence			Comments	Security Review	Occurrence			Comments
		Y	N	NA			Y	N	NA	
5.9	Has the customer at least three attempts to enter the PIN (i.e. limited number of login attempts)?				If the PIN is entered incorrectly several times in a row (usually three attempts per card insertion), the card will be retained as a security precaution to prevent an unauthorized user from discovering the PIN by guesswork. However, this reduces usability since users who have difficulties remembering their PIN might be locked out of the system.	Does the system is blocked after three PIN failed attempts to access the system?				—
5.10	If login fails (e.g. users forget their PIN) is there another authentication option to login to the MTM?				For security reasons, there are no other options for now but it should be provided to users.	If login fails, does the system avoid displaying what part of the authentication information was incorrect[30]?				The system must not provide any error feedback due to security reasons. However, the system needs update with current best practices on this matter.

Table 7.4 (Continued)   Usable Security Symmetry Inspection Method

#	Usability Review	Occurrence			Comments	Security Review	Occurrence			Comments
		Y	N	NA			Y	N	NA	
5.11	Have image quality, usability, and integrity (QUI) aspects been considered in check imaging according to the Check 21 Act regulations[31]?		(N)		• System review and implementation. • Image Quality (IQ) involves looking at the characteristics of the image (e.g, the size or the presence of all intact corners in a document. Image Usability (IU) involves looking at the legibility of the document to determine if the information on the check is readable and, hence, usable). Some of the common usability issues are: • Oversized documents can cause legibility and usability problems, depending on the degree of document overlap.	Does the system provide Image Integrity (II) in check imaging?		(N)		System review and implementation. Image Integrity (II) is to ensure that images and their associated metadata files match. If mismatches occur, financial institutions may face considerable risks with significant financial, customer trust, and public relations repercussions. In other words, the wrong image associated with transaction data can be a privacy issue if the wrong check image is displayed to the customer.

(Continued)

Table 7.4 (Continued)   Usable Security Symmetry Inspection Method

#	Usability Review	Occurrence			Comments	Security Review	Occurrence			Comments
		Y	N	NA			Y	N	NA	
5.11					• Dark streaks can result in the illegibility and/or potential non-usability of a required field within an image representation of an item. • "Spot noise" may negatively impact legibility and usability of the image. It is an isolated "small grouping" of connected black pixels surrounded on all sides by white pixels. • The appearance of "white streaks" may lead to image capture suspects and/or non-usability.					

(Continued)

Table 7.4 (Continued) Usable Security Symmetry Inspection Method

#	Usability Review	Occurrence			Comments	Security Review	Occurrence			Comments
		Y	N	NA			Y	N	NA	
5.12	Is Data Encryption Standard (DES) encryption avoided although it provides virtually no latency delays and minimal response time impact compared to Triple-DES (3DES)?				DES encryption is used though less secure. DES is a symmetric-key block cipher published by the National Institute of Standards and Technology (NIST) (i.e. commonly used method for encrypting data using a secret or private key).	Is Triple-DES (3DES)32 encryption employed, thus providing MTM transactions with greater security?				System review and implementation. Brute-Force Attacks Against DES: The most basic attack method is brute force. This consists of attempting every possible key against an encoded text. Since the overall key size for single DES is small at 56 bits, there have been successful brute-force attacks against the algorithm over the past two decades. As a result, any practical use of DES in industry today typically requires Triple DES in order to keep potentially sensitive information guarded against most attacks. Usability advantages of using 3DES correspond to the administrator, not end user, level as it is easier to implement and more secure, very efficient in hardware but not particularly in software. It is popular in financial systems, as well as for protecting biometric information.

(Continued)

Table 7.4 (Continued) Usable Security Symmetry Inspection Method

#	Usability Review	Occurrence			Comments	Security Review	Occurrence			Comments
		Y	N	NA			Y	N	NA	
										The current recommendation is to use Advanced Encryption Standard (AES), which is a specification for the encryption of electronic data established by NIST (U.S.) in 2001. It is based on a combination of a strong algorithm with a strong key. The Rijndael (its original name) block cipher can use different block and key lengths, such as 128-, 192-, and 256-bit. AES outperforms 3DES both in software and in hardware. High speed and low RAM requirements allow AES to perform well on a wide variety of hardware, from 8-bit smart cards to high-performance computers.

(Continued)

Table 7.4 (Continued)   Usable Security Symmetry Inspection Method

#	Usability Review	Occurrence			Comments	Security Review	Occurrence			Comments
		Y	N	NA			Y	N	NA	
5.13	Are event logs (i.e. process mining) analyzed and usability improvements implemented in the MTM system?				Process mining is a data mining technique for extracting useful knowledge from event logs in information systems. Event logs are the MTM transactions. An event log logs different events as they occur within a particular process of the system. Process mining discovers useful information regarding the process available in the event logs. Logging analysis is an efficient and effective method to investigate MTM usability.	Are event logs analyzed and security improvements and/or security compliance items implemented in the MTM system?				Logging must be used in full force because that is the easiest way to meet the PCI[33] requirements for specific events that need to be captured and retained.

(Continued)

Table 7.4 (Continued)   Usable Security Symmetry Inspection Method

#	Usability Review	Occurrence			Comments	Security Review	Occurrence			Comments
		Y	N	NA			Y	N	NA	
6. Usability Criterion: **Minimal Memory Load**										
Whether a user is required to keep a minimal amount of information in mind in order to achieve a specified task.										
6.1	Is the memory load on the user minimized?				This means that users do not need to memorize long data lists, complicated procedures, or undertake complex cognitive activities.	If the task at hand is complex, are the procedures or steps broken down into sub-steps to facilitate its understanding and execution in a secure way?				This refers to cognitive and physical requirements to perform a complex task. For example, transfer a large dollar amount to a checking account using KBA, OOBA, and palm recognition authentication methods). Out of band authentication (OOBA) is the use of two separate networks working simultaneously to authenticate a user. With OOBA, a transaction cannot complete without access to the second authentication network. This means using the mobile phone to verify the identity of the user involved in a local and web transaction at the MTM. The user will receive a call to authenticate the large dollar transaction before it completes.

(Continued)

Table 7.4 (Continued)   Usable Security Symmetry Inspection Method

#	Usability Review	Occurrence			Comments	Security Review	Occurrence			Comments
		Y	N	NA			Y	N	NA	
6.2	Has data entry been kept short to minimize input actions and memory load on the user?				Short-term memory (STM) capacity is usually limited to 7 ± 2 items (e.g. users might enter letters, digits, words, etc.), so the shorter the entries, the smaller errors and reading times. Data entry refers to user actions involving the input of data to a computer, and computer responses to such inputs.	Does the system provide displayed feedback for all user actions during data entry?				Usually, a system should provide feedback to users about the rejection of an entry. Also, data entries should be checked by the system for correct format, acceptable value, or range of values. When mnemonics (i.e. any learning technique that aids information retention in the human memory) or codes are used to shorten data entry, they should be distinctive and have a relationship or association to normal language or specific job-related terminology. Redundant entry, for instance, may be needed for resolving ambiguous entries, for user training, or for security (e.g. user identification). Finally, for reasons of data protection, it may not be desirable to display feedback for secure entries.

(Continued)

Table 7.4 (Continued) Usable Security Symmetry Inspection Method

#	Usability Review	Occurrence			Comments	Security Review	Occurrence			Comments
		Y	N	NA			Y	N	NA	
6.3	Are short PINs used, such as four digits or less?				Financial PINs are often four-digit numbers. Four-digit (or less) PIN number is easier to memorize and faster to type.	Is a PIN used in conjunction with a hardware device, providing stronger security (i.e. two-factor authentication?				This refers to two-factor authentication: something you *know* (i.e. PIN), and something you *have* (i.e. hardware token).
6.4	Are system-defined PINs avoided?				System-defined PINs are more difficult to memorize given that they have no meaning, cannot be pronounced, and are harder to remember than passwords. The recommendation from the usability point of view is to adopt a "user-defined PIN".	Can a system-defined PIN be combined with a variable, such as current date and time, at each login to eliminate the risk of replay? Does the system require an additional and faster authentication method to be used in conjunction with user selected PIN?				System review and implementation for both security reviews.

(Continued)

Table 7.4 (Continued) Usable Security Symmetry Inspection Method

#	Usability Review	Occurrence Y	N	NA	Comments	Security Review	Occurrence Y	N	NA	Comments
6.5	Is the MTM's application based on recognition of visual items for authentication (i.e. to avoid unaided recall)?		▓		System review and implementation. This would be a similar approach used by "SiteKey" from Bank of America, as already described in previous section.	If recognition of visual items for authentication is used, can the users also associate a phrase with an image to enhance security?			▓	System review and implementation.
6.6	Is the "time-to-learn" the silent alarm feature (almost) equal to zero so that users can intuitively send a silent alarm?	▓			Silent alarm is a safety PIN system that would alert police that a crime was in progress when a cardholder at an MTM keyed in the reverse of her PIN.	Is the alarm system reliable (i.e. nuisance alarms, false alarms, or failure to alarm when called for)?	▓			Nuisance alarms occur when an unintended event evokes an alarm status by an otherwise properly working alarm system.

(Continued)

Table 7.4 (Continued) Usable Security Symmetry Inspection Method

#	Usability Review	Occurrence			Comments	Security Review	Occurrence			Comments
		Y	N	NA			Y	N	NA	
6.7	Are the sequences and interdependencies of the MTM's artifacts and their corresponding UIs (i.e. business workflow) less harsh from the user interaction standpoint, thus enhancing customer intimacy?				Customer intimacy involves providing appropriate choices, information, and advice. There are some areas for improvement for this item. For example, the number of authentication methods should be reduced. This probably depends also on the interoperability of the system and applications.	Does the system's owner provide enhanced process control and operational risk mitigation regarding the MTM's business process?			—	

(Continued)

Table 7.4 (Continued) Usable Security Symmetry Inspection Method

#	Usability Review	Occurrence			Comments	Security Review	Occurrence			Comments
		Y	N	NA			Y	N	NA	
6.8	For PINs longer than four or five characters, are mnemonics or abbreviations being used?				Memorability is further improved by the use of mnemonics to aid their recall. The ISO 9564, an international standard for PIN management and security in retail banking, specifies that PINs shall be from four to twelve digits long, noting that longer PINs are more secure but harder to use. It also notes that not all systems support entry of PINs longer than six digits. The current MTM system does support the use of PINs longer than four digits.	For PINs and passwords, are mnemonics or abbreviations being avoided?				—

(Continued)

Table 7.4 (Continued)   Usable Security Symmetry Inspection Method

#	Usability Review	Occurrence			Comments	Security Review	Occurrence			Comments
		Y	N	NA			Y	N	NA	
7. Usability Criterion: **Resource Safety**										
Whether resources (including people) are handled properly, without any hazard.										
7.1	Does the system's owner place MTMs in branch lobbies, or other "indoor" locations such as supermarkets, etc.?	▓			"Indoor" MTMs are considered safer from the users' standpoint.	Apart from "indoor" locations does the system's owner provide MTMs that are inside locked, card accessible booths or rooms?		▓		System review and implementation.
7.2	Are users updated concerning of the owner's system security services to better meet customer needs?		▓		System review and implementation.	Is the dynamic loading of applications provided by the system?	▓			It refers to security upgrades and updates, as well as emergencies. Applications are loaded on an as-needed basis. The system needs to be able to add additional applications to the card platform.

(Continued)

Table 7.4 (Continued) Usable Security Symmetry Inspection Method

#	Usability Review	Occurrence			Comments	Security Review	Occurrence			Comments
		Y	N	NA			Y	N	NA	
7.3	If the MTM offers a silent alarm feature, is Reverse PIN used, providing faster recall?		▨		System review and implementation.	Has the system's owner security a policy clause that states that Reverse PIN must only be used for an emergency?	▨			System review and implementation.
7.4	When using a silent alarm, can users make use of the "spoken PIN" instead of the "typed" PIN?		▨		The "spoken PIN" is easier to trigger than the typed one, especially in emergency situations where users are under stress and pressure.	If the physical security of the customer is threatened (e.g, an attempt to steal a customer's card/money), can the user send a silent alarm to the owner's system?		▨		This silent alarm makes an audible noise elsewhere and notifies the system's owner (e.g. bank) and/or police. For example, a situation might be a customer who is forced by a robber to withdraw money from the MTM. So, s/he can type (or speak) an emergency PIN alerting the bank's security system.

(Continued)

Table 7.4 (Continued)   Usable Security Symmetry Inspection Method

#	Usability Review	Occurrence			Comments	Security Review	Occurrence			Comments
		Y	N	NA			Y	N	NA	
7.5	Is there any mechanism to avoid the card getting trapped inside the MTM's card reader (i.e. Lebanese loop)?				A Lebanese loop is a device used to commit fraud and identity theft by exploiting MTMs. It is a strip or sleeve of metal or plastic which blocks the MTM's card slot, causing any inserted card to be apparently retained by the machine, allowing it to be retrieved when the fraudster leaves.	Is there a way for users to avoid forgetting their cards on the card slot and preserve the security of the transaction?				With the new MTM card reader slot, users only need to insert the card in the slot and remove it quickly. The MTM does not keep the card inside the machine anymore during transaction processing, which increases security.
7.6	Are audible instructions (e.g. voice PIN, etc.) available so that people who have vision disability can use the machine?				A Talking MTM is a self-service machine that provides audible instructions to enable individuals who cannot read an ATM screen to independently use the machine. All private, spoken instructions are delivered through a standard headset that plugs into an audio jack on the machine front. The audio jack is located on the right-hand side of the MTM screen and is embossed with the sign of a headphone set.	Is PIN entered on public keypad then transmitted to the central server and checked against saved file avoided?				PIN is stored and checked against the smart card and not transmitted over the network where it would be susceptible to "sniffing" (i.e. shoulder surfing or direct observation by cameras). The risk of eavesdropping or replaying a PIN depends on how and where it is entered into the system. This option is not recommended from the security standpoint since the PIN has to travel over the network anyway.

(Continued)

Table 7.4 (Continued) Usable Security Symmetry Inspection Method

#	Usability Review	Occurrence			Comments	Security Review	Occurrence			Comments
		Y	N	NA			Y	N	NA	
					The talking functionality will be enabled when a headphone is detected. Also, an additional functionality enables users to utilize the machine with enlarged fonts, enabling users with low vision to operate the machine easily.					
7.7	Does the MTM hardware and aesthetics convey a secure and trusted operating machine to users?				According to Tractinsky et al. (2000)34, aesthetics issues could be related to usability. Aesthetics affected perceptions of usability more than actual usability. Aesthetics and usability are potentially independent dimensions, but both can contribute to user satisfaction.	Is the outside of the MTM tamper-evident (i.e. it makes unauthorized access to the protected object in the MTM easily detected)?				System review and implementation to keep up with new types of attacks.

(Continued)

Table 7.4 (Continued)   Usable Security Symmetry Inspection Method

#	Usability Review	Occurrence			Comments	Security Review	Occurrence			Comments
		Y	N	NA			Y	N	NA	
7.7					MTMs should be attractive. Potentially, good looking MTMs could attract new users and make existing users more comfortable. For instance, relaxing background music could relieve users anxiety during complex operations.					
7.8	Have users the ability to send a silent alarm through their mobile devices due to MTM's silent alarm failure?				System review and implementation.	Are attacks prevented on the mobile device that hosts the MTM's emergency application?				System review and implementation

(Continued)

Table 7.4 (Continued)   Usable Security Symmetry Inspection Method

#	Usability Review	Occurrence			Comments	Security Review	Occurrence			Comments
		Y	N	NA			Y	N	NA	
7.9	Is remote live assistance functionality available to users?				System review and implementation. For example, you can swipe your debit card, credit card, driver's license, or photo ID for authentication and then a live teller – physically based in another state – pops up on the screen to assist users. A member of the MTM technical support team can connect and control users' mobile devices and assist and solve problems remotely.	Are all live remote assistance sessions encrypted and password protected?				System review and implementation. How is security maintained? At the beginning of a screen-sharing session, users and the support expert are connected via a communication server. During the session, all transferred information, including screen views, file-transfer data, and identities, are encrypted. Encryption and decryption are from end to end, so data cannot be intercepted during transit. Also, the data is encrypted using 128-bit Advanced Encryption Standard (AES) encryption. After the session has ended, the support expert can no longer see your screen or access your computer unless you make another explicit request for support.

(Continued)

Table 7.4 (Continued)   Usable Security Symmetry Inspection Method

#	Usability Review	Occurrence			Comments	Security Review	Occurrence			Comments
		Y	N	NA			Y	N	NA	
7.10	Can users identify physically if keypads, card readers, etc. have been compromised by potential attackers?				• It is difficult to detect such compromises but some tips should be communicated clearly to users by the financial institution so that they can try to prevent such attacks when performing MTM transactions. Some of the tips are: • Look for anything out of the ordinary: a loose corner you can jiggle, something attached to the card slot, or a camera above the keypad that could take a picture of your pin. • Avoid ATMs that are away from busy areas or outside public vision. Hackers only target MTMs that are easy to access and modify without getting caught. • Hackers avoid MTMs from big banks and MTMs inside businesses.	Is the identity integrity of customers preserved (i.e. check the cardholder identity using, for example, an alternative identity verification method such as fingerprint, iris, retina, palm recognition, etc.)?				Fingerprint recognition is used. Criminals sometimes attach fake keypads or card readers to existing machines. This is an open door for Man-in-the-middle attacks which have been used to record customer's PINs and bank card information in order to gain unauthorized access to their accounts. Several MTM manufacturers have put in place countermeasures to protect the equipment they manufacture from these threats. Also, iris recognition is as good an option as fingerprint recognition in terms of usable security. It requires users to put their eyes in close proximity to the camera and is considered the most reliable biometrics. This method is non-contact, non-invasive, it can be located everywhere, and the physical attribute it uses is relatively stable.

Notes for Table 7.4

1 Multiple-technology enables a card to be produced with contact or contactless smart chip technology, magnetic stripe, bar codes, optical stripe, and/or 125 kHz proximity antenna. A card containing several types of read/write media is generally called a multiple technology card.

2 Contactless technology offers reliable and fast throughput. If another authentication factor is used, then the throughput advantages offered by contactless technology is decreased, but the strength of security and authentication are increased.

3 Physical access is being able to physically touch and interact with computers and network devices. Logical access refers to the connection of one device or system to another through the use of software. This software may run, say, as the result of a user powering a computer desktop, which then executes the login sequence, or it may be the result of internal processing between systems.

4 For example, different kinds of security measures are called for protecting information that is in transit (128-bit Advanced Encryption Standard (AES) encryption using an ephemeral encryption key that is destroyed when it is no longer used) versus information that is to be stored permanently on a hard drive (128-bit AES encryption for storing documents using key escrow, secret splitting, or even a secondary encryption key so that the document's content can be recovered in the event of a problem).

5 A cookie is a small piece of text stored on a user's computer by a web browser. It consists of one or more name-value pairs containing bits of information such as user preferences, shopping cart contents, identifier for a server-based session, or other data used by websites.

6 A single card solution for contact and contactless applications.

7 A microcontroller chip card (e.g. a card that opens a door by just bringing the card close to the card reader, and that same card also provides logical access to computers and networks).

8 Role-based access control (RBAC) is a general security model that simplifies administration by assigning roles to users and then assigning permissions to those roles.

9 It is important to understand the difference between customization and personalization. Customization refers to cases in which the user experience is adapted to each individual user's needs (e.g. a "portal" with headlines from the *Wall Street Journal*; enter ticker symbols for the stocks you want to track). Personalization is driven by the computer, which tries to serve up individualized pages to the user based on some form of model of that user's needs (e.g. Amazon "noticing" what you are buying and changing their front page to feature similar items).

10 Biometrics-Based Web Access experiments. Retrieved on January 3, 2009, http://citeseerx.ist.psu.edu/viewdoc/download?doi=10.1.1.27.7828&rep=rep1&type=pdf

11 FRR = False rejection rates: the likelihood that a legitimate person will be denied access; FAR = false acceptance rates: the likelihood that the wrong person will be able to access the system. The FRR ranges from 1 to 20% whereas the FAR ranges from 0,001 to 5%.

12 Users can authenticate to the portal, and then navigate to a PIN reset screen where the old PIN is required and validated using the rules set on the smart card during the chip personalization process.

[13] The estimated length of time that a system is available and operational between failures. It is a measure for hardware reliability.

[14] A bank's MTM sends a transaction again if the network went down before a confirmation message was received from the mainframe.

[15] The estimated length of time needed to bring a system back up and make it fully operational following a system failure.

[16] A bank's MTM sends a transaction again if the network went down before a confirmation message was received from the mainframe.

[17] It can be used to authenticate the identity of the cardholder by matching a live scan of a biometric feature (e.g. fingerprint or iris scan) to the template on the card.

[18] Consumers and Mobile Financial Services 2015 Report – March 2015, Board of Governors of the Federal Reserve System, Washington, DC. 20551. Consumer and Community Development Research Section of the Federal Reserve Board's Division of Consumer and Community Affairs (DCCA). http://www.federalreserve. gov/econresdata/consumers-and-mobile-financial-services-report-201503.pdf

[19] The card issuer uses the client registry to determine which applications are active and queries the application owner for the client backup database in the case of card replacement.

[20] For example, the contents of the physical and logical access control privileges databases can be combined into a single integrated database maintained as part of the card management system.

[21] This technique stores the enrollment template into the smart card's secure memory (i.e. the information is secured both during collection and during use of the credential in the ID system).

[22] Video encryption algorithms focus on protecting the more important parts of a video stream, thereby reducing the total amount of data encrypted and providing a faster response time which enhances the end-user experience.

[23] Not susceptible to "'sniffing" (i.e. a sniffing attack occurs when an attacker gains access to the network TCP/IP traffic path, captures data packets that make up the conversation, and assembles the packets into a format readable to the attacker). For example, the privacy of each user's credit card numbers is crucial. Although the Internet is by no means armoured, the most likely location for the loss of privacy to occur is at the last points of the transmission.

[24] Data corruption is the deterioration or damage of computer data caused by human, hardware, and software error.

[25] "Tamper-evidence for digital signature generation" can be used to defend against attacks aimed at covertly leaking secret information held by corrupted signing nodes. Retrieved on January 20, 2016, https://eprint.iacr.org/2005/147.pdf

[26] If a cardholder blocks their card by entering an invalid PIN, the cardholder should have the capability to unblock the card. When the cardholder's card is initially setup, a special unblock code should be generated, encrypted, and then stored in the card management system.

[27] Longer PINs give stronger security but bad usability because they are harder to remember and take longer to type.

[28] This would provide a higher level of security but lower system response time and thereby usability. The risk of eavesdropping the telephone network is a real threat, especially since it cannot be encrypted.

[29] Java Card Technology, Oracle. Retrieved on January 20th, 2016 http://www.oracle. com/us/technologies/java/embedded/card/overview/index.html

[30] To reduce the risk of successful intrusion attempts.

7.1.2.3.2.2 Severity Rates

Usability problems can significantly be eliminated or reduced through the severity rates, where evaluators are able to identify the security and usability problems that should be tackled and fixed. The ratings also aid in the allowance of resources for treating the user interface problems. If the severity ratings indicate that several usability problems remain in the UI, it will most likely be inadvisable to release it. However, one might choose to go forward with the release of a system with several usability problems if they are all judged as being cosmetic in nature.

According to Nielsen (1994a), the severity of a usability problem consists of a combination of three elements: frequency ranges (i.e. from ordinary problems to atypical ones), impact (i.e. establishes the ease or difficulty with which users recover from a problem), and persistence (i.e. ranges from just one problem that might be surmounted to a problem that continually replicates itself, becoming bothersome to users).

Last but not least, it is recommended to assess the *market impact* of the usability problem since certain usability problems can have a destructive effect on the popularity of a product, even if they are fairly easy to overcome.

After the symmetry inspection method has been performed by the evaluators, usability and security severity ratings are gathered in a matrix-based questionnaire. As shown in Table 7.5, these questionnaires contain the following information:

- Problem description, name of the criterion, severity rate, usability or security issue explanation, interdependencies, and recommendations.
- Usability and security problems that have been discovered.
- Severity rates of each problem which is specified by the evaluator.
- The descriptions are synthesized by the evaluator from the comments made for each problem. Typically, the evaluator need only spend about 30 minutes to give their severity ratings. The experience shows that severity ratings from a single evaluator are too unreliable to be trusted (Nielsen, 1994a). As more evaluators are asked to judge the severity of usability problems, the quality of the mean severity rating increases rapidly, and using the *mean of a set of ratings from three evaluators* is satisfactory for many practical purposes.
- The recommendations column suggests a solution to the problem.

■ **Usability Severity Ratings**

The 0 to 4 rating scale is employed to rate the severity of usability problems (Nielsen, 1994a). Table 7.6 shows an example of a usability problem and its severity rate:

- 0 = I don't agree that this is a usability problem at all.
- 1 = Cosmetic problem only: need not be fixed unless extra time is available on the project.

Table 7.5 Usability Severity Ratings and Recommendations for MTM

Problem Description	Criterion	Severity Rate	Usability Issue	Interdependencies	Recommendations
Vertical and horizontal scrolling is not possible in each window	Consistency and standards	1	Only two screens have scrollbars.	The speediness of the task is compromised.	Give user more control.

Table 7.6 Usability Severity Ratings and Recommendations for MTM

Problem Description	Criterion(a) Name	Severity Rate	Usability Issue	Interdependencies	Recommendations
No user personalization and customization options	Minimal Action and Operability	2	Users take more time to perform certain tasks on the MTM that would be otherwise performed faster and more efficiently. Examples are described in the Minimal Action and Operability criteria in the previous section, "Usable Security Symmetry Inspection Method". Using personalization and customization methods improves the efficiency and simplicity of usage of the MTMs and enables users to benefit from the machine according to their needs and in the shortest time duration and highest efficiency.	All MTM clients' records are stored (and retrieved) in the CRM system and other related databases. This data is gathered and updated at several touch points: • At the time of opening an account; • While the account is being used; • When customers demand during card usage. This data includes personal information, records of purchases, performed transactions such as time, place, or websites where those purchases have been done, people with whom customers have financial relations, type of the withdrawn sums, purchased tickets, payment invoices, and so on.	Provide user customization and personalization. Currently, customization and personalization capabilities are becoming best practices and even high-priority user experience and business requirements rather than just mere optional techniques to improve user satisfaction.

- 2 = Minor usability problem: fixing this should be given low priority.
- 3 = Major usability problem: important to fix, so should be given high priority.
- 4 = Usability catastrophe: imperative to fix this before the product can be released.

The usability severity rating for MTM featuring one particular problem description is presented in Table 7.6; however, detailed usability severity rating presenting more problem descriptions can be seen in Appendix 2.

■ Security Severity Ratings

The security severity ratings are based on six aspects:

- Authentication (i.e. user identity proofing and verification).
- Confidentiality (i.e. information is not made available or disclosed to unauthorized individuals, entities, or processes).
- Integrity (i.e. data has not been modified or destroyed in an unauthorized manner).
- Non-repudiation (i.e. the author of a document cannot later claim not to be the author; the "document" may be an email message, or a credit card order, or anything that might be sent over a network).
- Access control (i.e. granting access to data or performing an action; an authentication method is used to check a user login, then the access control mechanism grants and revokes privileges based on predefined rules).
- Availability (i.e. a computer system asset must be available to authorized parties when needed).

The 4 to 1 (Critical to Low) rating scale is employed to rate the severity of security problems. It has been developed by taking into account the following step: develop the Authentication Risk-assessment Matrix. The rating scale is described as follows:

- 4 = Critical Impact:
  - This rating is set to flaws that could be effortlessly exploited by a remote unauthenticated attacker and lead to system compromise (e.g. arbitrary code execution*) without involving user interaction. These categories of vulnerabilities can be exploited by worms. However, flaws that involve an authenticated remote user, a local user, or an improbable configuration would not be categorized as *critical impact*.
- 3 = Important Impact:
  - This rating is set to flaws that can effortlessly compromise the confidentiality, integrity, or availability of resources. These categories of

---

* Arbitrary code execution is employed to describe an attacker's ability to execute any commands of the attacker's choice on a target machine or in a target process.

vulnerabilities allow the following: local users to gain privileges, unauthenticated remote users to view resources that should be secured by authentication, authenticated remote users to perform arbitrary code, or local or remote users to effortlessly originate a denial-of-service (DoS).

- 2 = Moderate Impact:
  - This rating is set to flaws that might be harder or more unlikely to be exploitable but given the right conditions could still lead to some compromise of the confidentiality, integrity, or availability of resources. These categories of vulnerabilities are the ones that may well have had a *critical impact* or *important impact* but are less effortlessly exploited based on a technical evaluation of the flaw, or have an effect on improbable configurations.
- 1 = Low Impact:
  - This rating is set to all other issues that include a security impact. These categories of vulnerabilities require there to be improbable circumstances for them to be able to be exploited, or where a successful exploit would entail negligible consequences.

Table 7.7 presents an example of security problems related to usability criteria and their severity rates and recommendations featuring a single particular "Problem Description" (i.e., use case). A more complete coverage of use cases can be seen in Appendix 3.

7.2 Conclusion

As shown in this chapter, it is not enough to describe the task and elicit the conflict qualitatively (e.g. Chapter 6); it is crucial to also formulate those usability and security attributes into conflicts. This chapter presented a quantitative quantification of security and usability. Each of these two factors is decomposed into measurable criteria such as usability criteria and its corresponding usability and security severity ratings as shown in Tables 7.7 and 7.8).

This makes it possible to use usable security symmetry inspection for the requirements and design phase of the authentication method development lifecycle (AMDLC). It can be also used as an evaluation tool after the product (authentication method) has been released to manufacturing or market. In other words, our inspection method can be employed in two ways, such as *guiding the design decision* or *assessing the design of a final product* as follows.

- *Guiding the design decision (requirements and design phase)*: As already mentioned, by analyzing and answering concurrently the usability and security review questions, our inspection method forces user experience and/or security designers to think of the process as a whole (usability and security) rather than as part of the whole (usability or security). This method also forces them

Table 7.7 Example of a Security Problem and Its Usability Criterion and Severity Rate

Problem Description	Criterion	Severity Rate	Security Issue	Interdependencies	Recommendations
Machine requires only PIN and smart card when using common functionalities	Minimal Action	2	Machines requiring only PIN and smart card (two-factor authentication) have become pretty commonplace but also more exploitable by hackers. As technology advances, modern hackers are increasingly more malicious and savvy.	A two-factor authentication is not as strong for authentication as it was five years ago or even less.	Remove PIN as a requirement and implement a biometric authentication method (e.g. fingerprint) used in conjunction with a smart card. This will ease user interaction and improve security.

Table 7.8 Security Severity Ratings and Recommendations for MTM

Problem Description	Criterion(a) Name	Severity Rate	Security Issue	Interdependencies	Recommendations
Exposing security holes	Minimal Action	3	Using customization and personalization methods is one of the important methods for obtaining customer satisfaction. They improve the efficiency and simplicity of usage of the MTMs and enable users to benefit from the machine corresponding to their needs (e.g. service selection options, language, customized dashboard, etc.). However, to achieve the delivery of next-generation services via MTMs such as customization/ personalization, easy integration between end user and third-party capabilities is needed.	Third-party integration, and user customization and personalization.	1. Make a risk assessment to what degree of customization/personalization is realistic at the MTM without compromising security. 2. Use a hybrid contactless smart card to provide the required level of security for logical access, while providing a reliable and easy-to-use solution for physical access. Or use multiple technology cards that can combine either of the ISO/IEC standard contactless smart card technologies with 125 kHz proximity technology. This enables the card to operate with legacy physical access control systems, as well as new ISO/IEC-compliant systems. Providing multiple read/write capabilities on a card can often assist in providing the tools needed to enable a transition from legacy to new technology applications over time. 3. Make a risk assessment to what degree of customization/personalization is realistic at the MTM without compromising security.
			Although distributed hardware and software architecture can deliver a fully customizable service, this may open the door for security holes (e.g. JavaScript that runs in the MTM's browser can be exploited to hack accounts of users who visit a particular user profile page).		4. Organizations should take advantage of the card architecture by linking physical and logical access privileges to increase security within the card.

to initiate – or trigger – potential solutions in their minds when working on the review questions. Finally, another important aspect is that while evaluating the review questions they will be able to anticipate the identification of potential bugs earlier in the requirements and design phase that would otherwise occur when the product is handed off to manufacturing.

■ *Assessing the design of a final product (readiness phase)*: The method can also be applied to *assess the design of a final product* (i.e. authentication method) which has been already released to manufacturing or to the market. The method is used in this case in the *readiness* phase of the AMDLC. It can be used to point out what security and usability fixes and/or improvements should be implemented in the next release of the authentication method. This is a very important aspect of the AMDLC which can represent benefits in terms of improved product reliability, reduced customer support calls, smaller releases product lifecycle, reduced number of bugs to be opened against the product, and, finally, enhanced user interaction (i.e. finding bugs earlier can force designers to review and improve authentication method functionalities with the development and product management teams).

Chapter 8

The Usable Security Protocol Methodology: Demonstrate

8.1 Introduction

The verification and validation (V&V) phase of the Multifunction Teller Machine (MTM) case is undertaken through a demonstrational approach. V&V are supplementary techniques aimed at checking the quality of the system generated. Verification is a quality control process that is employed to assess whether or not a product, service, or system conforms to regulations, specifications, or conditions included at the start of a development phase. Verification is frequently an internal process.

Validation is a quality assurance process of laying down facts that provide a high degree of assurance that a product, service, or system accomplishes its planned requirements. The crucial goal of validation is to make the model useful; it addresses the right problem and provides correct information about the system being modeled.

Typically, verification can be expressed by the question "Are we building the thing right?" and validation by "Are we building the right thing?". The latter refers back to the user's needs, while the former "building it right" checks that the specifications are being properly implemented by the system.

As we will be demonstrating our usable security protocol, first let us remember what an experiment is for. In scientific investigation, an experiment is a method of investigating causal relationships among variables. An experiment is a foundation of the empirical approach to acquiring data about the world and is employed in both natural sciences and social sciences. An experiment can be employed to help in resolving a practical problem which is the focus of this book: the usable security of user authentication methods.

Security systems should be viewed as socio-technical systems that depend on the social context in which they are embedded to function correctly. Security systems will only be able to offer the intended protection when people truly understand and are able to use them correctly. As Jøsang et al. (2007b) stress, there are genuine differences between the degree by which systems can be considered theoretically secure (supposing those systems are properly operated) and in reality secure (acknowledging that frequently those systems will be operated erroneously). Often, as already stated, there is a trade-off between usability and theoretical security. It can be useful to reduce the level of theoretical security to improve the whole level of actual security. For example, the strongest passwords are the ones, from a theoretical perspective, randomly generated. However, given that it is difficult to remember such passwords, people will write them down, and in so doing they will weaken the system's security. Thus, it might be important to let people choose passwords that are easier to remember. Even though this reduces the theoretical strength of the passwords, it intensifies the security of the system as a whole.

To this end, to illustrate our demonstrational approach, we make use of the usable security inspection method to evaluate a one-time password (OTP) authentication.

8.1.1 The Demonstration of One-Time Password Authentication

This section describes the demonstration of the OTP authentication method which has been subjected to the GOMS analysis. The goal of the OTP demo is to show the difficulties users are subject to when using this particular method. It focuses on the usability aspects of user interaction with the system. This demo is defined as a wireless- and-token-based authentication task which is comprised of the following elements:

- Wireless local area network (WLAN).
- Hardware token with OTP functionality.
- Personal identification number (PIN).
- Tokencode.

8.1.1.1 Wireless Local Area Network (WLAN)

A wireless network refers to any type of computer network that is associated with a telecommunications network whose interconnections between nodes are

Figure 8.1   RSA SecurID® 700 authenticator.

implemented without the use of wires. Wireless networks are generally implemented with some type of remote information transmission system that uses electromagnetic waves, such as radio waves, for the carrier, and this implementation usually takes place at the physical level or "layer" of the network, as shown in the Open System Interconnection (OSI) model*.

More specifically, a WLAN is a wireless distribution method for two or more devices that use high-frequency radio waves and often include an access point to the Internet. A WLAN allows users to move around the coverage area, often a home or small office, while maintaining a network connection.

8.1.1.2 Hardware Token With OTP Functionality

The RSA SecurID® 700 authenticator† (hardware token) (Figure 8.1) is an OTP scheme and authentication mechanism to protect an organization's most critical information assets. It is generally used to secure either local or remote access to computer networks. Each end user is assigned an authenticator that generates a one-time-use code (tokencode). When logging on, users simply enter this tokencode plus their standard PIN (passcode) to be successfully authenticated.

RSA SecurID® 700 is a widely accepted authentication scheme used by many infrastructure solutions. It is a battery-powered handheld device containing a dedicated microcontroller. The microcontroller stores the current time and a 64-bit seed value that is unique to a particular token in the random-access memory (RAM). When the token is manufactured, a seed is encoded into the specific token. RSA SecurID® 700 comes pre-seeded and is ready-to-use out-of-the-box. Used in conjunction with RSA® Authentication Manager 7.1‡, the RSA SecurID® 700 adds an additional layer of security by requiring users to identify themselves with two unique factors – something they know, a PIN, and something they have, a unique

---

* Open System Interconnection (OSI) protocol. February, 2016 <http://docwiki.cisco.com/wiki/Open_System_Interconnection_Protocols>
† RSA SecurID® 700 authenticator. February, 2016 <http://www.emc.com/collateral/datasheet/rsa-secureid-700-authenticator-ds.pdf>
‡ RSA® Authentication Manager 7.1. February, 2016 <http://theether.net/download/RSA/SecurID/7.1SP2/auth_manager_administrator_guide.pdf>

OTP that changes every 60 seconds – before they are granted access to the secured application.

An authentication server verifies the passcode. The server maintains a database which contains the seed value for each token and the PIN or password for each user. From this information, and the current time, the server generates a set of valid passcodes for users and checks each one against the entered value. The PIN can be changed if forgotten. The OTP is the concatenation of the four-digit user PIN and the six-digit tokencode. Patented RSA technology synchronizes each authenticator with the security server, ensuring a high level of security.

The OTP, something you *have*, is coupled with a secret PIN, something you *know*, to generate a combination that is almost impossible for a hacker to guess. Due to the algorithm being secret and only known to a limited number of individuals, attacks against the process of generating new token codes are less likely to succeed, unless information about it leaks to the public or a hacker deliberately reverse-engineers the hardware token and the authentication server. Other attacks, such as network traffic sniffing, shoulder surfing, keyboard logging, and social engineering, might have limited success. If by one of these methods a hacker manages to obtain a tokencode and knows the user's PIN number, the hacker can only use the passcode to authenticate within a limited period of time – up to 60 seconds after the tokencode was initially generated on the RSA SecurID® 700 token. Moreover, users must not have used this tokencode to authenticate or it will be rejected as already used. If the hacker does not try authentication within the 60-second period after having obtained actual credentials, the chance is lost and the hacker needs to find another tokencode. However, if a hacker manages to capture a legitimate tokencode from a user, the hacker already has the PIN as the first four digits of the passcode. The PIN can be used for brute-force attacks, decreasing the unknown keyspace from 10^{10} to 10^6 but this is still not likely to lead to success due to the limited time that the hacker has. Beyond the 60-second validity period of the tokencode, other attacks such as password guessing, sniffing, man-in-the-middle, etc. are not possible because the tokencode has already changed. RSA SecurID® 700 is also tamper-evident, meaning that if someone opened the token for criminal purposes, it would be apparent to the user of the device.

8.1.1.3 Personal Identification Number (PIN)

A PIN is a secret numeric password shared between users and systems, and it can be used to authenticate users to systems. A PIN is usually a four-digit number, though sometimes it can have six digits or more. Users are required to provide a non-confidential user identifier or token and a confidential PIN to gain access to the system. Upon receiving the User ID and PIN, the system looks up the PIN

Figure 8.2   Tokencode: a six-digit number.

based upon the User ID and compares the looked-up PIN with the received PIN. Users are granted access only when the number entered matches with the number stored in the system.

8.1.1.4 Tokencode

The hardware token displays a new pseudo-random value, usually a six-digit number, called the tokencode at a fixed time interval, usually 60 seconds, as shown in Figure 8.2.

8.1.2 How the OTP Demonstration Works

To access resources protected by the RSA SecurID® 700, users combine their secret PIN with the tokencode generated by their authenticators. The outcome is a unique, one-time-use passcode that is used to positively identify and authenticate users. If the passcode is validated by the RSA SecurID® 700 system, users are granted access to protected resources. If it is not recognized, users are denied access.

■ **System Requirements**
  A virtual private network (VPN) client application should be installed in the user's laptop which is an EMC VPN Client Version 4.8.02.0010 2006. Client type: Windows, WinNT.
■ **Demonstration Steps**
  Eleven demonstration steps of the OTP authentication method using the EMC VPN client are described in the following:
  – **Step 1:** User accesses the EMC VPN application by clicking on the yellow unlock padlock icon in the laptop taskbar:

  – **Step 2:** System opens up the EMC VPN application on the screen by displaying a connections entry list for authentication:

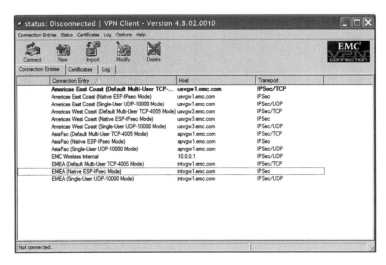

- **Step 3:** User selects and double-clicks the connection entry for his/her area:

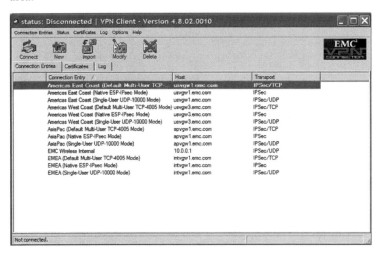

- **Step 4:** System initializes the authentication process by contacting the authentication server which displays the login screen window;

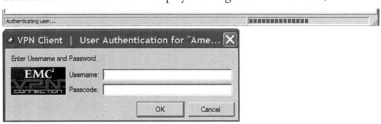

– **Step 5:** User enters his/her username in the username field:

– **Step 6:** User enters his/her PIN number in the passcode field. The PIN number is provided when user enrolls in the RSA® Authentication Manager 7.1 application:

– **Step 7:** User refers to the RSA SecurID® 700 hardware token by reading and memorizing the tokencode displayed in the token's liquid-crystal display (LCD). The system generates a different tokencode every 60 seconds.
– **Step 8:** User concatenates the tokencode (six-digit number) to the PIN already entered in the passcode field and clicks on the "OK" button:

– **Step 9:** System contacts the security gateway;

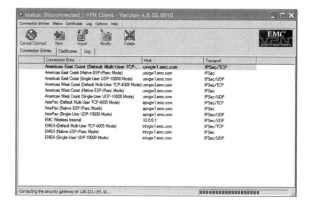

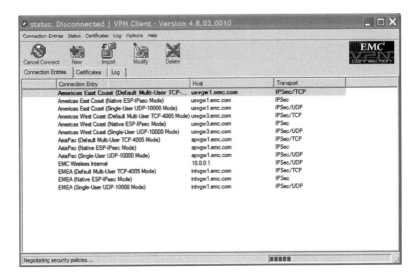

– **Step 10:** User clicks on the "Continue" button:

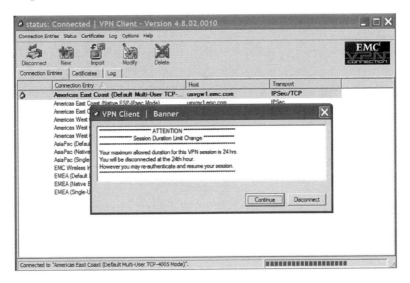

– Detail of the pop-up screen:

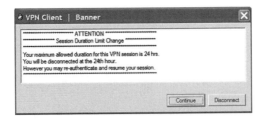

- **Step 11:** If authentication is successful, the system grants access to the user. The yellow padlock icon that is shown as locked in the taskbar means that the authentication was successful. If authentication was not successful, the system prompts the user to enter login details again.

8.1.3 One-Time Password Usability Testing

To inspect the usability of the OTP authentication method using the RSA® SecurID 700 hardware token, a usability testing was performed at RSA Security (EMC) in Bedford, Massachusetts on June 16, 2009.

Confidentiality Disclaimer: To protect RSA Security confidential information, certain usability testing questionnaires and tests results regarding this OTP usability testing cannot be disclosed.

8.1.3.1 Terms and Definitions

- **Correlation**: Correlation shows that as one variable increases, the other decreases. This also shows that there is no connection between the two variables. For example, as temperature increases, the sales of ice-cream are also likely to increase. Therefore, there is likely to be a positive correlation.
- **Data point**: A data point is a discrete unit of information. In a general sense, any single fact is a data point. In a statistical or analytical context, a data point is usually derived from a measurement or research and can be represented numerically and/or graphically.
- **Invisible user authentication (IUA)**: IUA involves actively introducing additional identifiers with the simple addition of a cookie and/or a flash shared object (i.e. flash cookie) which can then serve as a more unique identifier of a user's device. It can also track characteristics that are a natural part of any device such as HTTP headers, operating system versions, browser version, languages, and time zone. IUA also uses behavioral profiling to track user behavior. This involves identifying the activities that are performed by the device and the user and matching that data against the historical profile of activity related to that user to determine if there is inconsistent or unusual behavior that may indicate unauthorized access.
- **One-time password**: An OTP is a password that is valid for only one login session or transaction on a computer system or other digital device. OTPs avoid a number of shortcomings that are associated with traditional (static) password-based authentication; a number of implementations also

incorporate two-factor authentication by ensuring that the OTP requires access to something a person has (such as a small keyring fob device with the OTP calculator built into it, or a smart card or specific mobile phone) as well as something a person knows (such as a PIN).

■ **RSA Federated Identity Manager (FIM):** RSA FIM enables end users to collaborate with partners, service providers, and multiple offices or agencies – all with a single identity and logon. Web single sign-on can provide a better user experience by allowing users to sign into a web portal just once.

■ **RSA SecurID® Appliance:** It is a core component of the RSA's Identity Assurance portfolio, delivers RSA® Authentication Manager, the software engine behind the industry-leading multi-factor user authentication technology, in an integrated, rack-mountable hardware appliance. Used in combination with RSA SecurID authenticators and contextual Risk-Based Authentication, the RSA SecurID Appliance validates the identities of users before granting access to critical company resources. Additionally, the system logs all transactions and user activity, allowing administrators to use it as an auditing, accounting, and compliance tool.

■ **Security Socket Layer (SSL):** SSL is the standard security technology for establishing an encrypted link between a web server and a browser. This link ensures that all data passed between the web server and browsers remain private and integral.

■ **Virtual private network:** A VPN extends a private network across a public network such as the Internet. It enables users to send and receive data across shared or public networks as if their computing devices were directly connected to the private network and thus are benefiting from the functionality, security, and management policies of the private network.

8.1.3.2 Objectives Of The OTP Usability Testing

Identify high-priority usability issues:

■ Assess the usability of designs for end user tasks involving IUA-enabled SSL-VPN.
■ Assess the usability of designs for end user tasks involving account creation and management using the RSA® Credential Manager* Self-Service Console, a web-based solution for the deployment and lifecycle management of RSA SecurID® authenticators.

---

* RSA® User Credential Manager. February, 2016 <http://www.emc.com/collateral/data-sheet/h11915-ds-rsa-sid-credential-manager.pdf>

8.1.3.3 Testing Tools

The following testing tools have been used in the usability testing:

- **Medium fidelity clickable prototype:** A visual guide that represents the skeletal framework of our application. Usually lacks typographic style, color, or graphics. The main focus lies in functionality, behavior, and priority of content.
- **Morae recording software*.** Morae allows the gaining of valuable insight about a product and removes the guesswork from decisions. We can record user interactions, analyze the results, and instantly share them with anyone – all within Morae and for any type of research.
- **System usability survey (SUS):** This survey provides a "quick and dirty", reliable tool for measuring the usability. It consists of a ten-item questionnaire with five response options for respondents; from "Strongly agree" to "Strongly disagree". Originally created by John Brooke in 1986, it allows us to evaluate a wide variety of products and services, including hardware, software, mobile devices, websites, and applications.
- **Additional surveys to gauge ease of use:** This might be internal and external off-the-shelf surveys that the authors consider reliable sources.

8.1.3.4 Testing Session

Session length: Approximately one (1) hour.

Number of external participants: Twelve (12).

Recruitment: It was centered on non-admin users with average computer skills and familiarity with remote login procedures.

Users were categorized into three levels:

- **Low:** Consistent computer user, not familiar with the concept of VPNs.
- **Medium:** Moderately skilled computer user, moderate access to remote resources but does not have in-depth knowledge or understanding of the agent or VPN.
- **Advanced:** Skilled computer user, frequently accesses remote resources, knowledgeable about VPNs.

8.1.3.5 Testing Methods: Participant Tasks

Participants were divided into two groups:

---

* MORAE recording software, TechSmith Corporation, February, 2016 <https://www.techsmith.com/morae-features.html>

- **Group 1:**
  - Authenticated via an IUA-enabled SSL-VPN.
  - Configured challenge method (i.e. security questions).
  - Logged into the self-service console.
  - Requested RSA SecurID® 800 hybrid authenticator.
  - Enabled token.
- **Group 2:**
  - Created a self-service account.
  - Requested IUA enablement.
  - Configured challenge method (i.e. on-demand tokencode).
  - Authenticated via IUA-enabled SSL-VPN.

8.1.4 Data Results

- System Usability Scale:
  - What does it measure?
    - Effectiveness (can users successfully achieve their objectives).
    - Efficiency (how much effort and resource is expended in achieving those objectives).
    - Satisfaction (was the experience satisfactory).
      - Average Score = 72.5 (scale of 1–100).
  - a. Historical testing results:

    a. RSA FIM SUS = 75.00
    b. RSA SecurID® Appliance Round 1 SUS = 67.92
    c. RSA SecurID® Appliance Round 2 SUS = 81.92

  - Areas that tested well:
    - Consistency.
    - Perception of the ability to learn the system quickly.
  - Areas of improvement:
    - Complexity of the system.
    - Need for prior knowledge.
  - Interesting data point – When rating the system on "cumbersomeness", Group 1 gave it the overall worst rating and Group 2 gave it the overall best of all the survey questions.
  - Most tasks were completed with high success rates, but some did need additional help from the facilitator.
  - There wasn't a strong correlation between user experience level and ability to complete most tasks easily.
    - The one exception was configuring on-demand as the challenge method.
    - All other tasks were well tolerated at all user levels.

8.1.5 Findings Summary

- Strengths:
  - Two-page logon process in the SSL-VPN.
  - Use of context-sensitive help (i.e. What is this? link): Context-sensitive help is a category of online help that is obtained from a specific point in the state of the software, providing help for the situation that is associated with that state.
  - Save this device.
  - Security questions:
    - Concept.
    - Specific questions available.
  - SSL page designs.
  - "Request a token" self-service console pages.
- Areas of Improvement:
  - Concept and terminology of on-demand service and tokencodes.

8.2 One-Time Password Usability Issues: Discussion

This section brings up and discusses usability issues regarding the RSA SecurID® 700 hardware token, which is in this book subjected to a demonstration test. The authors' comments are shown between parentheses (e.g. (Authors)) in the sub-sections.

Usability has been always a concern when designing an authentication method according to the main authenticator manufacturers such as RSA Security (EMC), VeriSign, and Vasco. However, achieving actual usability is, in fact, a most difficult task when taking into consideration user interaction.

RSA Security manufactures the RSA SecurID® 700 and claims the following usability characteristics of its hardware token:

8.2.1 Convenient Form Factor

> "With its robust keyring, small size and easy-to-read liquid-crystal display (LCD), the SecurID 700 is a convenient form factor for employees, partners, and customers."

(Authors) The LCD is not easy to read since, due to its design and form, shadow and light reflect off the display, making it difficult to read the passcode from the LCD. Actually, the user has to turn the token in several different positions until the passcode is readable. The token is, in fact, a bit big, not small. What was small five years ago (or less) is in fact big nowadays. For example, the very first mobile phones were really big and clunky; then they were small enough to fit in your briefcase, then your pocket, and then the palm of your hand. Due to the growing trends toward miniaturization (e.g. nanotechnology) and portability, users will therefore have more artifacts that can fit anywhere and are easier to carry, which reflects on the usability aspect of the product itself.

The token fits to some extent on a ring of keys, but because it is quite big it makes the fob too heavy. Also, to put the token on the fob ring is difficult; the ring is too thick so you have to use a knife or scissors in order to slide the token ring through the fob ring. Finally, it doesn't fit well in a pocket, because, as already mentioned, the token is quite big.

> "Users can easily read the OTP displayed on the authenticator* and know when the number is going to change by watching the countdown indicator."

(Authors) The LCD is not easy to read according to the previous comment.

8.2.2 Reliable Authentication Solution

> "The SecurID 700 authenticator is designed to withstand the worst imaginable conditions, offering industry-leading reliability. From temperature cycling to mechanical shocks to being immersed in water, the SecurID 700 is subjected to rigorous tests to ensure that customers do not face hidden costs due to token failures. The combination of this high level of quality with a lifetime warranty allows organizations to reduce the overhead costs of distributing replacement tokens and drive down the overall cost of security while providing a consistent and easy-to-use authentication experience for end-users[†]."

(Authors) An OTP is still a difficult method to handle in user authentication for the vast majority of average computer users. Users have difficulty understanding the concept of concatenation[‡]; it means users have to first input their PIN (a four-digit number) then concatenate the passcode (six-digit number generated by the

---

[*] In "Convenient Form Factor" section. RSA-The Security Division of EMC. February 24, 2016 <https://www.emc.com/collateral/software/data-sheet/h9061-rsa-securid.pdf>

[†] In "Reliable Authentication" section. RSA-The Security Division of EMC. February 24, 2016 <https://www.emc.com/collateral/software/data-sheet/h9061-rsa-securid.pdf>

[‡] The "Usability Severity Ratings and Recommendations" and "Security Severity Ratings and Recommendations" artifacts, evaluators can propose new solutions to ease the user authentication task. Also, a comparative analysis could be undertaken by identifying, on one hand, the potential cognitive problems (e.g. complex concatenation of PIN and tokencode) and, on the other hand, potential supplementary actions (e.g. filling out an additional field beyond username and passcode); but this is outside the scope of this book.

server) displayed on the hardware token LCD. An OTP authentication method requires users to utilize different artifacts and different communication channels, such as hardware tokens, web interfaces, laptops, and wireless networks.

The OTP method violates 22 usability criteria within the usable security symmetry inspection method* as described in the following. It concerns the perceptual and cognitive workload for individual inputs or outputs which represent the following usability review questions:

- 8-out-of-22 usability review questions specified in the *minimal action* criterion;
- 3-out-of-7 usability review questions specified in the *minimal memory load* criterion;
- 2-out-of-7 usability review questions specified in the *resource safety* criterion;
- 2-out-of-12 usability review questions specified in the *load time* criterion;
- 5-out-of-19 usability review questions specified in the *operability* criterion;
- 2-out-of-10 usability review questions specified in the *security* criterion.

A description of each usability criterion and its corresponding usability review questions are given in the following sections:

8.2.2.1 Usability Criterion: Minimal Action

The capability of the application to help users achieve their tasks in a minimum number of steps (i.e. the length of transactions and procedures).

- Is the workload low and simple (e.g. input workload kept to a minimum)?
- Is the authentication (verify/authenticate the identity of the user) process simple for users?
- Does the user have to authenticate using different communications channels?
- For codes longer than four or five characters, are mnemonics or abbreviations being used?
- Has the requirement of data entry by the user been limited when the data can be derived by the application?
- For data entry, has the system displayed currently defined default values in its appropriate data fields?
- Does the system accommodate both experienced and average computer users (e.g. length of transactions; shortcuts available to experienced users)?

---

* It is difficult for the concept of concatenation PIN + passcode to be understood by users, and it can be tested not only employing the usable security symmetry inspection method but also usability test (i.e. observing users performing the OTP task). Other usability inspection techniques can be used such as pluralistic walkthrough, task analysis, etc.

■ If users must switch between different systems, are dual-interface chip smart cards used?

8.2.2.2 Usability Criterion: Minimal Memory Load

Whether a user is required to keep a minimal amount of information in mind in order to achieve a specified task.

■ Are the sequences and interdependencies of the MTM's artifacts and their corresponding UIs (i.e. streamlined business workflow) less harsh in the user interaction viewpoint, thus enhancing customer intimacy (providing appropriate choices, information, and advice)?
■ Is the memory load on the user minimized (i.e. no memorization of long data lists, complicated procedures, or undertake complex cognitive activities)?
■ Are the entries short (i.e. short-term memory (STM) capacity is limited[43], so the shorter the entries, the smaller errors and reading times)?

8.2.2.3 Usability Criterion: Resource Safety

Whether resources (including people) are handled properly, without any hazard.

■ Are audible instructions (or voice PIN) available so that people who cannot read an MTM screen can independently use the machine?
■ If the MTM offers a silent alarm feature, is Reverse PIN used providing faster recall?

8.2.2.4 Usability Criterion: Load Time

The time required for the application to load (i.e. how fast it responds to the user).

■ If login fails, is there another authentication option available to users in order for them complete their task?
■ Does the live image require a small memory allocation which will affect the time for retrieval of the image if solicited by the user?

8.2.2.5 Usability Criterion: Operability

Amount of effort necessary to operate and control an application.

■ Is the user task workload light?
■ Can users customize the user interface to their specific needs (e.g. personalized look and feel)?

■ Can (some) system's mechanisms be configured by users to operate in a certain way (e.g. customer's language and "favorite transaction")?

■ When biometrics is not available, is there an alternative authentication method in order to provide availability to the users?

■ If users forget their PINs, can users reset them via a web interface rather than in an issuance station (i.e. ensure customer convenience & satisfaction)?

8.2.2.6 Usability Criterion: Security

The capability of the application to protect information and data so that unauthorized persons or systems cannot read or modify them and authorized persons or systems are not denied access.

■ Do users have more than a few alternatives to authenticate to the system to improve the availability and convenience of the system?

■ For logical access application, is biometrics provided to enhance usability (i.e. on-card biometric match and on-card key generation)?

Appendix 1: Authentication Risk-Assessment Matrix

Authentication Asset/Target	Threat (T) Description	Vulnerability (V) Description	CIA	Threat Rating	V Rating	Overall PCE OP = T + V	Asset Value Classification	Asset Value Exposure	Overall Exposure/Impact TI = AVC + AEC	RA TR = OP × TI	Risk Reduction Strategy
1. PASSWORD Personally identifiable medical information stored on *Structured Query Language* (SQL) Server.	Account credentials of data entry clerk stolen.	Overly complex password requirements cause users to write down passwords and leave them in obvious places.	CIA	3 Medium	3 Medium	6	4 Substantial	4 Serious	8	48	Mitigate by reducing password complexity requirements, enforcing policy to not leave passwords in obvious places, and providing user training in password use. Compromise of SQL data could result in large fines as a result of HIPAA[1] violations. Also, loss of public confidence could result in long-term loss of business.
2. PASSWORD Personally identifiable medical information stored on SQL Server.	Help desk resets password for an account used by a data entry clerk based on a request from an unauthorized individual (Social Engineering attack[2]).	Lack of policies and procedures in place to verify the identity of the individual requesting password reset.	CIA	2 Low	2 Medium	4	4 Substantial	4 Serious	8	24	Mitigate by implementing tighter procedures to verify identity and by delegating password change permissions to small business units where requests for password changes come from individuals known to the administrator. Compromise of SQL data could result in large fines as a result of HIPAA violations. Also, loss of public confidence could result in long-term loss of business.
3. PASSWORD Non-proprietary and non-confidential company data.	Summer intern's password guessed by attacker.	Use of familiar names and words in passwords.	CIA	1 Very Low	3 Medium	4	1 Negligible	1 Negligible	2	8	Mitigate by providing user education to all staff, including temporary staff. Compromised data results in minimal loss of productivity, etc. Costs of loss are easily absorbed.

(Continued)

Authentication Asset/Target	Threat (T) Description	Vulnerability (V) Description	CIA	Threat Rating	Overall PCIE		Overall Exposure/Impact			RA	Risk Reduction Strategy
					V Rating	$OP = T + V$	Asset Value Classification	Asset Value Exposure	$TI = AVC + AEC$	$TR = OP \times TI$	
4. PASSWORD Microsoft Exchange Hosted Services (EHS).	Attacker runs an online brute-force attack to determine the password of the account used for EHS and causes account lockout on the service account.	Account lockout thresholds set on domain. All versions of UNIX/Linux/Mac OS Server may be affected by accounts having weak or dictionary-based passwords for authentication. Most Unix/Linux systems include multiple standard services in their default installation. All versions of Unix/Linux/Mac OS Server are potentially at risk from improper and default configurations.	A	2 Low	3 Medium	5	3 Medium	3 Moderate	6	30	1. Disable account lockout thresholds on domain accounts, set them to a higher value, or set a low value for the lockout duration. Loss of EHS could result in loss of productivity and affect many users. 2. Don't use default passwords on any account. 3. Enforce a strong password policy. Don't permit weak passwords or passwords based on dictionary words to resist brute-force attacks. Use public key authentication mechanism for SSH[3] to thwart such attacks. 4. Limit the number of failed login attempts to exposed services. 5. Limit the accounts that can log in over the network; root should not be one of them. 6. Prohibit shared accounts and don't use generic account names like tester, guest, etc. 7. Log failed login attempts. A large number of failed logins to a system may require a further check on the system to see if it has been compromised. 8. Consider using certificate based authentication. 9. If your Unix system allows the use of Pluggable Authentication Modules (PAM), implement those that check the password's strength. Avoid service interactions and misconfigurations. Where possible, limit the functions of the host.[4] 10. Misconfigurations in multiple services may often increase the risk to a service.

(Continued)

Authentication Asset/Target	Threat (T) Description	Vulnerability (V) Description	CIA	Threat Rating	Overall PC/E		Overall Exposure/Impact			RA	Risk Reduction Strategy
					V Rating	OP = T + V	Asset Value Classification	Asset Value Exposure	TI = AVC + AEC	TR = OP × TI	
5. PASSWORD Company data residing on individual workstations of employees in the Human Resources department.	Attacker runs a brute-force attack against the local Security Accounts Manager[5] (SAM) of workstation in an attempt to acquire account credentials of the local administrator account.	Account lockout thresholds not set on member workstations.	CIA	2 Low	3 Medium	5	3 Moderate	4 Serious	7	35	1. Administrator User Account: The password should be unique and complex. For example, mitigate by requiring account lockout thresholds on workstations used by human resource employees. Loss of confidentiality of HR data could subject the company to severe legal sanctions and other consequences. 2. Guest User Account: This account is disabled by default and should remain configured that way. 3. New User Accounts: There should be few, if any, additional user accounts in the local SAM. 4. Administrators Group: The membership should be limited. 5. Account Policies: It is best to have the local SAM account policy meet or exceed the company security policy setting for the domain account policy. 6. User Rights: The user rights on every server should be audited to ensure that they are configured properly and no additional users have privileged access to the server. 7. Audit Policy: An important setting for both servers and clients to keep track of when users logon and what resources are accessed and when.

(Continued)

Authentication Asset/Target	Threat (T) Description	Vulnerability (V) Description	CIA	Threat Rating	Overall PC/E V Rating	Overall PC/E OP = T + V	Overall Exposure/Impact Asset Value Classification	Overall Exposure/Impact Asset Value Exposure	TI = AVC + AEC	RA TR = OP × TI	Risk Reduction Strategy
6. PASSWORD Network Infrastructure.	Account credentials of administrator stolen by individual "shoulder surfing" administrator performing daily tasks.	Lack of multi-factor authentication. Lack of secure work area for administrators.	CIA	2 Low	5	7	5 High	5 Severe	10	70	Mitigate by requiring administrators to use multi-factor authentication (e.g. smart card plus Personal Identification Number [PIN]), by providing secure work area for administrators, and by providing training to administrators regarding the use of credentials when others are able to view them. Theft of administrative credentials could have a severe impact throughout the company.
7. PASSWORD Random company data throughout organization.	Attacker exploits older authentication mechanisms.	Windows 95 and 98 clients using LAN Manager[6] (LM). It uses a particularly weak method of hashing a user's password known as the LM hash algorithm.	CIA	2 Low	3 Medium	5	4 Substantial	4 Serious	8	40	Avoid by installing the Active Directory Services client on Windows 9x workstations and disabling LM authentication throughout the domain. Possibility for extensive losses of confidentiality; however, the actual total impact rating depends on the actual account and data compromised.

(Continued)

Authentication Asset/Target	Threat (T) Description	Vulnerability (V) Description	CIA	Threat Rating	Overall PCE		Overall Exposure/Impact			RA	Risk Reduction Strategy
					V Rating	OP = T + V	Asset Value Classification	Asset Value Exposure	TI = AVC + AEC	TR = OP × TI	
8. PASSWORD Horizontal privilege escalation.	Potential web application vulnerabilities that may lead to this condition include easily guessable passwords.	Consider this scenario: Alice has access to her bank account in an Internet Banking application. Bob has access to his bank account in the same Internet Banking application. The vulnerability occurs when Alice is able to access Bob's bank account by performing some sort of malicious activity.	CIA	2 Low	3 Medium	5	4 Substantial	4 Serious	8	40	1. Remove unused services/apps from network devices. 2. Assign and enforce a strict password policy (e.g. strong password, changed regularly) and ensure all device access is password protected. 3. Disable unused system accounts and remove unused user accounts. 4. Tighten privilege levels on apps and storage locations; do not accept defaults, such as windows, which allows "everyone" access by default. 5. Regularly study log files and baseline system files for later comparison for tampering.
9. PASSWORD	Allowing password aging ✪	Allowing password aging to occur unchecked can result in the possibility of diminished password integrity.	CIA	1 Very Low	3 Medium	4	1 Negligible	1 Negligible	2	8	The recommendation that users change their passwords regularly and do not reuse passwords is universal among security experts. In order to enforce this, it is useful to have a mechanism that notifies users when passwords are considered old and that requests that they replace them with new, strong passwords. In order for this functionality to be useful, however, it must be accompanied with documentation which stresses how important this practice is and which makes the entire process as simple as possible for the user.

(Continued)

Authentication Asset/Target	Threat (T) Description	Vulnerability (V) Description	CIA	Threat Rating	Overall PC/E		Overall Exposure/Impact			RA	Risk Reduction Strategy
					V Rating	$OP = T + V$	Asset Value Classification	Asset Value Exposure	$TI = AVC + AEC$	$TR = OP \times TI$	
10. PASSWORD	Buffer overflows in the username or password of a login feature. ✿	Buffer overflow using long strings of "A" characters in username/ password during authentication. Web application components (e.g. CGI, libraries, drivers, web application server) in some languages that do not properly validate input can be crashed and used to take control of a process.	CIA	3 Medium	3 Medium	4	5 High	5 Severe	9	36	Keeping systems up-to-date with the most current security patches and using, for example, McAfee Entercept[7] will protect servers against these powerful threats.
11. PASSWORD	SQL Injection using "'" in the Username/ password of an authentication routine; AND "ID" or other identifier field. ✿	Attackers that can compromise passwords, usernames, or database fields can defeat authentication and/or have access to sensitive data.	CIA	4 High	4 High	8	4 Substantial	4	8	64	Developers need to either stop writing dynamic queries and/or prevent user supplied input, which contain malicious SQL, from affecting the logic of the executed query. There are several techniques for preventing SQL Injection vulnerabilities (OWASP_09).

(Continued)

Authentication Asset/Target	Threat (T) Description	Vulnerability (V) Description	CIA	Threat Rating	V Rating	Overall PC/E OP = T + V	Asset Value Classification	Asset Value Exposure	Overall Exposure/Impact TI = AVC + AEC	RA TR = OP × TI	Risk Reduction Strategy
12. PASSWORD Hard-coded or undocumented account/password	Insiders with privileged password access. ❂	Client-side systems with hard-coded passwords	CIA	2 Low	2 Low	4	3 Medium	3 Medium	6	24	Users should not circumvent password entry with auto logon, application remembering, embedded scripts, or hard-coded passwords in client software for systems that process/store mission critical and/or confidential data. Users should always deny having a password "remembered."
13. PASSWORD Authentication Tokens	Broken authentication and session management	Account credentials and session tokens are not properly protected. Attackers that can compromise passwords, keys, session cookies, or other tokens can defeat authentication restrictions and assume other users' identities.	CIA	3 Medium	3 Medium	6	3 Medium	3 Medium	6 M	36	Unless all authentication credentials and session identifiers are protected with Secure Socket Layer (SSL)[8] at all times and protected against disclosure from other flaws, such as cross site scripting, an attacker can hijack a user's session and assume their identity.

(Continued)

Authentication Asset/Target	Threat (T) Description	Vulnerability (V) Description	CIA	Threat Rating	Overall PCE		Overall Exposure/Impact			RA	Risk Reduction Strategy
					V Rating	OP = T + V	Asset Value Classification	Asset Value Exposure	TI = AVC + AEC	TR = OP × TI	
14. PASSWORD Random company data throughout the network.	Attacker intercepts network authentication packets that contain password hashes and attempts to break them offline.	NTLM[9] and LM[10] authentication used on the network.	CIA	2 Low	3 Medium	5	4 Substantial	4	8	40	Mitigate by upgrading clients to Windows 2000 or better. Possibility for extensive losses of confidentiality; however, actual total impact rating depends on the actual account and data compromised (It can also be mitigated by securing physical infrastructure of network cabling). One of the reasons for the relatively high value for the overall risk is that it is relatively easy to perform this exploit using downloadable tools.

(Continued)

Authentication Asset/Target	Threat (T) Description	Vulnerability (V) Description	CIA	Threat Rating	V Rating	OP = T + V	Asset Value Classification	Asset Value Exposure	TI = AVC + AEC	TR = OP × TI	Risk Reduction Strategy
					Overall PC/E		**Overall Exposure/Impact**			**RA**	
15. OUT OF BOUND AUTHENTICATION (OOBA) Trojan horse (TH)	Sending out spam email inviting users to access a website that will install a smart Trojan on a user's client computer.	The TH will observe activities on the client computer and get into action when, for example, the user starts an online banking session. When the user specifies a funds transfer transaction, the TH will alter the amount and destination account without displaying the alteration on the screen. The online bank will thus receive a transaction request with the false amount and destination account. Even when the transaction requires authorization via an SMS message, a significant percentage of users will fail to notice that the transaction details have been altered.	CIA	2 Low	3 Medium	5	3 Substantial	3 Moderate	6	30	This type of attack is difficult to detect. A strong network security policy with no unauthorized downloads is usually the best way to defend against Trojan horses. Also, perform random file comparisons of key binaries on hosts to known, good binaries, confirming that key binaries haven't been compromised.

Notes for Appendix 1

[1] The Health Insurance Portability and Accountability Act of 1996 (HIPAA) Privacy Rule http://www.hhs.gov/ocr/privacy/hipaa/understanding/index.html

[2] A social engineering attack is one in which the intended victim is somehow tricked into doing what the attacker requests.

[3] Secure Shell or SSH is a network protocol that allows data to be exchanged using a secure channel between two networked devices. Used primarily on Linux and UNIX based systems to access shell accounts.

[4] A host, network host, or Internet host is a computer connected to the Internet.

[5] The Security Account Manager (SAM) is a database present on servers running Windows Server 2003 that stores user accounts and security descriptors for users on the local computer.

[6] A network operating system from Microsoft that runs as a server application under OS/2. It supports DOS, Windows, and OS/2 clients. LAN Manager was superseded by Windows NT Server, and many parts of LAN Manager are used in Windows NT and 2000.

[7] The Entercept Management System delivers comprehensive, enterprise-class management for Entercept's intrusion prevention agents. McAfee Secure Computing http://www.mcafee.com/us/resources/data-sheets/ds-host-intrusion-for-desktop.pdf

[8] Secure Socket Layer (SSL) is a protocol developed by Netscape for transmitting private documents via the Internet. SSL uses a cryptographic system that uses two keys to encrypt data – a public key known to everyone and a private or secret key known only to the recipient of the message.

[9] NTLM is a suite of authentication and session security protocols used in various Microsoft network protocol implementations. NTLM authentication is a challenge-response scheme, consisting of three messages, commonly referred to as Type 1 (negotiation), Type 2 (challenge), and Type 3 (authentication).

[10] The LM authentication protocol, also known as LAN Manager and LANMAN, was invented by IBM and used extensively by Microsoft operating systems prior to NT 4.0. It uses a password encrypting technology that is now considered insecure.

Appendix 2: Usability Severity Ratings and Recommendations for MTM

Problem Description	Criterion(a) Name	Severity Rate	Usability Issue	Interdependencies	Recommendations
1. No user personalization and customization options	Minimal Action and Operability	2	Users take more time to perform certain tasks on the MTM that would be otherwise performed faster and more efficiently. Examples are described in the Minimal Action and Operability criteria in the previous section, "Usable Security Symmetry Inspection Method." Using personalization and customization methods improves the efficiency and simplicity of usage of the MTMs and enable users to benefit from the machine according to their needs and in the shortest time duration and highest efficiency.	All of an MTMs clients' records are stored (and retrieved) in the CRM system and other related databases. These data are gathered and updated in several touch points: • At the time of opening an account; • While the account is being used; or • Customers demands during card usage. These data include personal information, records of purchases, performed transactions such as time, place, or websites where those purchases have been done, people with whom customers have financial relations, type of the withdrawn sums, purchased tickets, payment invoices, and so on.	Provide user customization and personalization. Currently, customization and personalization capabilities are becoming best practices and even high-priority user experience and business requirements rather than just mere optional techniques to improve user satisfaction.
2. Heavy authentication workload	Minimal Action	3	Heavy user interaction workflow: Users are required to log into the system through different types of authentication methods and levels of authentication according to the type of services they are trying to access. These activities are in fact difficult and cumbersome for end users.	System performance: Perception of performance is informed by two factors: the speed with which an application processes data and performs operations and the speed with which the application responds to users.	Single Sign-On (SSO) authentication across multiple service providers: SSO removes the need for authenticating users at every instance of remote service provision. However, this involves not only the system's owner but also third-party players, so an agreement should be settled to provide such functionality to users.

(Continued)

Problem Description	Criterion(a) Name	Severity Rate	Usability Issue	Interdependencies	Recommendations
3. Heavy re-authentication workload	Minimal Action	2	Tiresome re-authentication to users: For security reasons, an MTM implements an authentication time limit after a certain period of inactivity. These time limits may cause problems for certain people (especially with disabilities) because it may take longer for them to complete the activity. When users are logged out while still in the midst of a transaction, it is important that they be given the ability to re-authenticate and continue with the transaction without the loss of any data already entered. Also, due to the atomicity, consistency, isolation, and durability (ACID) properties, if users have entered data during any workflow and are logged off of the MTM, the system will not save data so that it cannot be used after a user re-authenticates. ACID has four primary attributes ensured to any transaction by a transaction manager. For example, if users want to transfer some amount of money from one account to another, then users would start a procedure to do it. However, if a failure occurs, then, due to atomicity, the amount will either be transferred completely or will not be even initiated. Thus, atomicity protects the user from losing money due to a failed transaction.	Authentication session time limit and process management such as ACID, where every transaction must conform strictly to the ACID properties.	Longer authentication session time limit. System review and implementation.

(Continued)

Problem Description	Criterion(a) Name	Severity Rate	Usability Issue	Interdependencies	Recommendations
4. Increased number of user logins	Minimal Action	2	Using different credentials on different systems decreases frequency of use and memorability: Credentials might be passwords, PINs, OTPs, biometrics, etc. while different systems can be mobile network, web-based apps, MTM software (e.g. Windows XP professional or embedded, Windows 10, Linux, Android), etc.	Distinct login methods for individual and particular tasks.	Reducing the number of login workflows will create a more efficient and streamlined workflow environment for users. Social Login Workflow: It enables users to use their existing accounts over different social networks and identity providers, such as Facebook, Google, Twitter, Yahoo, LinkedIn and others to login or register on a system. Single Sign-On (SSO): It can be a solution for individual systems which tend to store user credentials locally. Care must be taken with this kind of approach since administration, as well as usability, issues will arise if this method is used in a network environment.
5. Matters of software and system inconvenience	Minimal Action	2	Users must switch between different systems, such as mobile network, MTM software (e.g. Windows 10), web-based apps (e.g. buy a concert ticket in an e-commerce site).	Lack of motivation to explore other functions within the application might compromise task completion, such as buying a concert ticket.	Identifying clusters of similar, related, or high-priority workflows may reduce *the number of workflows* and provide a more efficient and streamlined system environment for users.

(Continued)

Problem Description	Criterion(a) Name	Severity Rate	Usability Issue	Interdependencies	Recommendations
6. Lack of user personalization (authentication)	Operability	2	Users cannot choose their preferred authentication method (system-defined). Too many steps to login with certain authentication methods (e.g. OTP), and certain web apps that require users to create an account (pre-registration) before they can access a website.	Control, trust, and "explorability."	Streamline the authentication process by providing authentication methods more convenient to users. Control, trust, and "explorability" are fundamentally important to any system. If users feel in control of the process, they will be more comfortable using the system. If users are comfortable and in control, they will trust that the system will protect them from making unrecoverable or unrecognized errors or from feeling stupid.

(Continued)

Problem Description	Criterion(a) Name	Severity Rate	Usability Issue	Interdependencies	Recommendations
7. Physical and logical access inconvenience	Minimal Action	3	Users need to carry different cards for different accesses to apps or services.	Productivity for corporate users, partial fulfilling or task abandonment for consumers (e.g. buying a concert ticket).	For physical and logical access applications, contactless technology provides quick user throughput (i.e. the number of users that can be processed (or authenticated) per unit time for a given system). Also, to accommodate the user's need for a single ID credential, using a contactless card for both physical and logical access could be attractive. Depending on system requirements, a contactless smart card can be used to provide the required level of security for logical access while providing a reliable and easy-to-use solution. Contactless technology has the advantage of not suffering from physical contact contamination or requiring precise insertion and release.[1]

(*Continued*)

Problem Description	Criterion(a) Name	Severity Rate	Usability Issue	Interdependencies	Recommendations
8. Lack of user interface customization	Operability	2	New software is continually released with larger numbers of features and more complex UIs and application interoperability. As the interfaces of software grow, the user's efficacy when working with the MTM apps diminishes.	Mostly, customers make use of some of the tools and navigation menus available at the MTM. All extra information displayed in the UI that are not essential for the accomplishment of their tasks becomes superfluous and prone to distraction, which potentially slows down their workflow.	Adaptable interfaces, or user-controlled interfaces, provide mechanisms with which users can customize the interface themselves. Another alternative is a hybrid approach with a variety of mixed-initiative interfaces, where control over the adaptation process is shared between the system and the user. Also, this customization should be direct and simple, that changes would be local as well as global, and that the system would be able to support deep customization (Stuerzlinger et al., 2006) (i.e. users are able to create new sets of options and change the function of tools).
9. Excessive and varying levels of access and different networks	Operability	2	Users get irritated when they attempt to access services and get interrupted to log in again. Security is a secondary task for users. To perform some transactions, users can get very confused due to the number of authentication methods needed and different levels of physical and logical access for different services.	Technology and access to different networks local area network (LAN), wide area network (WAN), or Internet (web apps) when performing certain transactions.	SSO can be the solution.

(Continued)

Problem Description	Criterion(a) Name	Severity Rate	Usability Issue	Interdependencies	Recommendations
10. Unbalanced FRR and FAR of hand geometry recognition	Operability	2	Legitimate users may get irritated if the system doesn't accept them, and they may go elsewhere or abandon the task.	Throughput performance.	The accuracy of each system can be adjusted. However, there are some drawbacks to doing so (e.g. changing the system sensitivity to reduce false rejections may increase false acceptances). Zang (2004) has documented the hand geometry recognition system scanners as being able to operate with a low FAR of 0.096% and reasonable FRR of 1.05%.
11. Locked out user: forgotten PIN.	Operability	5	To request a new PIN or change a PIN when users don't know their PIN, they need to call their financial institution, go in person to a financial center, or request it by mail. The financial institution produces a system-generated PIN upon users' request and mails it to them. This security policy represents inconvenience and dissatisfaction for users who need to access their account at that very moment at the MTM.	Unavailability of services.	To request a new PIN or change a PIN when users don't know their PIN, an option is an on-demand authenticator which grants "emergency" access through a hardware or software token. By successfully completing life question challenges, users can keep making transactions by requesting on-demand authentication even in off-hour scenarios. To change a PIN when users do know their PIN, provide an option in the MTM navigation menu: users select "More Options" then "Change PIN". Then they will be able to accomplish their task and goal.

(Continued)

Problem Description	Criterion(a) Name	Severity Rate	Usability Issue	Interdependencies	Recommendations
12. Inefficient mean time between failures (MTBF): Hardware reliability	Operability	3	The MTM system is unavailable at certain very short periods of time. Users might try to log in to the system at that exact moment. There is no way of estimating how long the failure could last and what exactly can cause a failure. Failures might occur from the network for a number of reasons.	Unavailability of services.	The system needs to be updated regarding technology infrastructure. High availability – often implemented with failover clustering, load-balanced clustering, warm standby servers, and log shipping.
13. High rate of mean time to repair (MTTR)	Operability	3	System reliability, which affects authentication tasks and availability of services to users.	Unavailability of services.	The system must have an MTTR of zero, which means that it has redundant components which can take over the instant the primary one fails.
14. High total transaction time (3Ts) for biometrics	Operability	2	User throughput is too high for a biometric authentication method. Users may feel uncomfortable using the system because it takes too much time to capture their samples (satisfaction) and to learn the interface (learnability). In addition, users' workflow time is increased, which prevents them from using the product or service quickly (efficiency). As already mentioned, the 3Ts corresponds to the time for a single user to present the biometric (acquisition time) and processing time.	Task incompletion, unavailability of services, and loss of productivity for corporate users.	Although certain transactions that use biometric authentication method take 15 seconds, which might be acceptable in some environments, this could be less than ideal for processing a large number of users in the MTM case. 3Ts should be adjusted to a minimum of four and a maximum of ten seconds for a single user. Hand geometry uses low computational cost algorithms, which leads to fast results.

(Continued)

Problem Description	*Criterion(a) Name*	*Severity Rate*	*Usability Issue*	*Interdependencies*	*Recommendations*
15. Hidden privacy policy	Privacy	2	The privacy policy is located under the "Security" link in the MTM's Welcome screen.	Liability.	Provide a "Privacy Policy" link which directs users to the privacy policy in the Welcome screen.
16. No user control over private user information	Privacy	3	Personal information is precious to many Internet businesses. Users are unaware that they are not in control of their private information. Increasing practice of commercializing privacy by publicly businesses progressively creates new risks for users in return for little to no protection or reward. User privacy should be taken very seriously if an organization wants to be credible to its customers.	Liability and business best practices, ethics, and standards.	Adopt privacy best practices in the MTM network. The use of comprehensive safeguards to protect the privacy and confidentiality of users' information. The main items for a privacy policy are: • Privacy Highlights: Provide highlights of the full Privacy Code and apply to personal information collected, used and disclosed. • Privacy Code: Inform users of the ways the financial institution helps protect their privacy and the confidentiality of their information.

(Continued)

Problem Description	Criterion(a) Name	Severity Rate	Usability Issue	Interdependencies	Recommendations
					• Online Privacy Code: Show how the financial institution handles the information they collect when customers use its MTM network and websites. • Privacy Agreement: Show how the financial institution collects, uses and discloses users' personal information and outlines their options to withdraw or refuse their consent.

(Continued)

Problem Description	Criterion(a) Name	Severity Rate	Usability Issue	Interdependencies	Recommendations
17. No cardless transactions available to users	Operability	2	Inconvenience to users. Currently, the MTM only accepts transactions from customers using a card. This can be inconvenient for users due to the following: • Funds transfer is very easy and secured • When you need cash urgently but you have your MTM card or you have forgotten the PIN. • Transaction can be done 24/7. • Recipient is not required to have bank account, which is the most distinguishing feature of this service. • Users have to carry more cards in their wallets.	System and services availability.	Cardless new technology allows customers to withdraw money from cash machines using their mobile phones. Users are given a six-digit code to allow them to enter into an MTM to release the cash. This is actually a long-predicted move toward the smartphone becoming a digital wallet.
18. Inappropriate performance of encryption algorithm	Operability	2	System response time is slow when users access video at the MTM. If the appropriate encryption policy is not used, then the user experience and performance will suffer.	Productivity for corporate users, partial fulfilling or abandonment of a task.	One solution, for example, is to use video encryption algorithms which focus on protecting the more important parts of a video stream, thereby reducing the total amount of data encrypted and providing a faster response time which enhances the user experience.

(Continued)

Problem Description	Criterion(a) Name	Severity Rate	Usability Issue	Interdependencies	Recommendations
19. Nonexistent authentication method option	Load Time	2	If the PIN is forgotten, there is no other authentication method option to login into the system. Unfortunately, this is a hurdle that causes most users to end up quitting on achieving their goals.	Availability of services, and task abandonment.	The system should provide an authentication screen that prompts them to reset their PINs.
20. User insecurity (sensitive information)	Privacy	2	Loss, misuse, modification, or unauthorized accesses to sensitive information (e.g. social security numbers, credit card numbers, driver license numbers, etc.) negatively affect the privacy of users depending on the level of sensitivity and nature of the information.	Higher authentication abandonment rates, partial fulfilling, task abandonment, or even in-person visit to a branch instead of making a transaction on the MTM, which from users' perspective is more secure.	• Use strong authentication procedures and other access controls to make information usable by authorized individuals. • Reduce the volume of collected and retained information to the minimum necessary. • Limit access to only those individuals who must have such access. • Encrypt data. • Give users control on how their personal data is used by third parties.

(Continued)

Problem Description	Criterion(a) Name	Severity Rate	Usability Issue	Interdependencies	Recommendations
21. Failing and annoying PIN login	Security	3	If users forget their PIN (after inserting their cards), the system blocks users after three failed attempts to log on. This is extremely annoying for users, and it is even more aggravating if they are in a hurry or have an emergency.	Availability of services.	If a cardholder blocks their card by entering an invalid PIN, the cardholder should have the ability to unblock the card. When the cardholder's card is initially set up, a special unblock code should be generated, encrypted, and stored in the card management system.
22. Failing and annoying KBA login over the telephone	Security	3	When trying to login using KBA over the phone, users need to provide a PIN as a second authentication factor (strong authentication), but if they forget their PIN, they will need to go through other authentication methods in order to provide that second factor. This is difficult for users who will need to reset all their credentials which can compromise them to finish their tasks.	Availability of services.	Most users are much better remembering four-digit random PINs than any other type of random password. If users forget their PIN, provide them with an emergency method such as security questions (or other alternatives) to replace temporarily their PIN. Additionally, users must ensure their telephone banking PIN is not similar to their four-digit PIN used for MTM transactions.

(Continued)

Problem Description	Criterion(a) Name	Severity Rate	Usability Issue	Interdependencies	Recommendations
23. Slow and unprotected transaction	Security	2	Inconvenience to users when on-card biometric match is not provided. There is a separate physical access control system database, managed by the facilities organization, which maintains an employee's physical access control privileges and issues the proximity card.	System performance, interoperability, and scalability.	Biometrics can be used with card technologies (e.g. smart cards), where biometric information is stored on the card and then verified with the received biometric at the point of interaction. On-card biometric match is the concept of either matching or storing, for instance, a fingerprint on a smart card, and there is no need for a database. Matching fingerprint information in the card removes the uncertainty of matching on a network-connected device, external server, or a database. This provides faster transactions (performance), user acceptance and more security. On-biometric card match technology is a good candidate to replace PINs and passwords, irrespective of the card technology or the application.

(Continued)

Problem Description	Criterion(a) Name	Severity Rate	Usability Issue	Interdependencies	Recommendations
24. Difficulty coping with different communications channels.	Security	2	Overwhelm customers with complexity, heavy flow of interaction, and memorability issues when dealing with different services and different communication channels, such as MTM, web, and mobile network. Although it might be considered a convenient service when one does not have access physically to an MTM, it does place the burden on users with regard to the coordination of the MTM with a mobile phone. However, users are still required to authenticate to the system by entering a PIN. Unlike passwords, PINs have no meaning and are even harder to remember for users.	Performance and availability of services.	Minimize interaction while keeping the most important services always available. Reduce the number of actions required to perform the required tasks. Provide other services by flagging as optional or customizable services but maintain a collection of "core services."

(Continued)

Problem Description	Criterion(a) Name	Severity Rate	Usability Issue	Interdependencies	Recommendations
25. Slow response time	Load Time	2	People are impatient at the MTM; they are in a hurry, they want to get things done, and they do not want to waste their time. Anything that slows them down will frustrate them. Slow response time can distract users in their interaction with the machine. This especially annoying and difficult when they are trying to accomplish main transactions such as money transfer, cash withdrawal, reset a PIN, etc.	Performance, availability of services, partial fulfilling or task abandonment.	Java Card is a robust multifunction card containing a cryptographic processor and secure storage token offering a solid platform on which to securely store a biometric template (e.g. digital hand or fingerprint) and to execute an on-card biometric-match function. Higher-end microprocessor cards include a Central Processing Unit (CPU)[2] for performing computations on locally stored data. Finally, Java Card technology preserves many of the benefits of the Java programming language – productivity, security, robustness, tools, and portability – while enabling Java technology for use on smart cards.

(*Continued*)

Problem Description	Criterion(a) Name	Severity Rate	Usability Issue	Interdependencies	Recommendations
					Some measures to help in the mentioned interdependencies are: Speed of retrieval: Only essential information should be displayed. Images usually cause the MTM's system to download slower. • Number of lines of code: Lines of code indicates the approximate number of lines in the code. A very high count might indicate that a type or method is trying to do too much work and should be split up. It might also indicate that the type or method might be hard to maintain and might run slow, which can harm the system performance (e.g. long style sheets, too many scripts, etc.).

(Continued)

Problem Description	Criterion(a) Name	Severity Rate	Usability Issue	Interdependencies	Recommendations
					• Appropriate balance between speed and design for usability. Standard guidelines for response times from Nielsen (1993) still apply nowadays. There are three main time limits (which are determined by human perceptual abilities) to keep in mind when optimizing web and application performance: *0.1 seconds* (one-tenth of a second): ideal response time. The user doesn't sense any interruption; *1 second*: highest acceptable response time. Download times above 1 second interrupt the user experience; *10 seconds*: unacceptable response time. User experience is interrupted at an alarmingly high rate and users are likely to abandon the system.

(Continued)

Problem Description	Criterion(a) Name	Severity Rate	Usability Issue	Interdependencies	Recommendations
26. Nonexistent image quality and image usability (check imaging)	Load Time	2	Poor check quality images bring exposure to fraud, liability, and system trust issues. This will negatively impact customer confidence. Check Image Quality (IQ), Image Usability (IU), and Image Integrity (II) are modules that provide solutions for identifying substandard check image quality, unusable fields, or images which mismatch the Magnetic Ink Character Recognition (MICR)[3] index. The capture system resident in a wide range of workflows can ensure that the check image captured meets the quality and warranty requirements for image exchange and Check 21.	Availability of services, task abandonment, and system trust.	Adopt check recognition software that performs a wide range of image quality, usability, and integrity tests on each image and "flags" those items which represent a quality, usability, or negotiability risk for financial institutions. It addresses image quality concerns on both a per-image and scanner level. Through this approach, a financial institution has the ability to review and correct image quality and usability issues before they become concerns. Assurance ensures the images for forward presentment meet the image quality and usability definitions.

(Continued)

Problem Description	Criterion(a) Name	Severity Rate	Usability Issue	Interdependencies	Recommendations
27. User privacy and integration	Load Time	2	As already mentioned, DES is not suitable for confidential data and usability. This relates to the administrator and end user levels. Some of the usability aspects are: • 3DES was the answer to many of the limitations of DES. Since it is based on the DES, it is very easy to modify existing software to use 3DES. Nowadays, the majority of ATMs do adopt 3DES. • Advanced Encryption Standard (AES)[5] is the substitute for DES (NIST, 2009) but 3DES will be maintained for *compatibility reasons* for several years after that. • AES is at least as strong as 3DES and probably *much faster*.	Availability of services, interoperability, and encryption algorithms.	According to the National Institute of Standards and Technology (NIST), after December 31, 2015, the use of AES-128, AES-192, AES-256, and three-key Triple DES is "Acceptable."[4] 3DES is an outstanding and reliable choice for the security needs of highly sensitive information, including PIN (if DES is used to encrypt PIN, the PIN is vulnerable to attack). Finally, to improve the transmission speed of 3DES, there is technology available today that enables it to reduce backup sizes and speed the transmission of the customer's backup via the Internet. Encryption algorithms notes.[6]

(Continued)

Problem Description	Criterion(a) Name	Severity Rate	Usability Issue	Interdependencies	Recommendations
28. Extensive automatic audit logs[7]	Load Time	2	Extensive automatic logs take excessive processing time, which lowers system performance and frustrates users. Automatic logs are used with software-based products to keep a record of all user interactions and transactions, but they can have the disadvantage that they will also pick up accidental actions.	Availability of services, system performance, task abandonment.	Typically, the MTM system implements extensive auditing through the various applications that comprise the system. Such logs are collected at three distinct levels: operating system, database, and application. First, it is likely that the logging is turned on full force because that is the easiest way to meet the PCI requirements for specific events that need to be captured and retained. Second, the recommendation is to do some research as to what can be done in terms of the configuration of the ATM system to pare down the high volume of traffic (i.e. configure specific types of audit logs the machine should record, not automatic logs). Strike a balance between automatic audit logs and particular configured audit logs (e.g. transaction logs).

(Continued)

Problem Description	Criterion(a) Name	Severity Rate	Usability Issue	Interdependencies	Recommendations
29. Lack of communication on security updates	Resource Safety	2	Users can get surprised if, for example, a new emergency access authentication method has been introduced in the system and there was no prior and/or clear communication about it. Users may feel insecure or unsure if they are performing transactions correctly in relation to any security aspects. Good communication, especially related to security, is crucial toward customers to promote trust and trustworthiness in MTM systems.	Availability of services; users are not confident about the MTM system; discomfort.	Provide automatic updates pushed to the machine. Send out notifications, alerts, updates concerning security to customers by mail or online and provide a link in the Welcome page to the "Security Update Services" section.

(Continued)

Problem Description	Criterion(a) Name	Severity Rate	Usability Issue	Interdependencies	Recommendations
30. No Reverse PIN available	Resource Safety	4	Customer is able to send a silent alarm using a Reverse PIN in response to a threat at the MTM and get help from the bank. Users have difficulty remembering PINs, especially under pressure.	User physical security, system trust.	The adoption of a Reverse PIN for emergency helps users in remembering it (e.g. PIN is 2637, users enter 7362). This would avoid having a new emergency PIN issued and the bothersome task of having to remember another one. *MTM Reverse PIN*: After the user enters the correct PIN in reverse order, their account is automatically locked and the system triggers an error message on the screen saying that the user's account has been locked. At the same time, the system also dials the preconfigured National Security Force number to have the robber trapped. This is clearly a resource safety feature that MTM offers to its customers. To unlock their account, the user needs to call customer service and provide, for example, answers to security questions as an additional authentication mechanism, and/or other private information stored in the system owner's database.

(Continued)

Problem Description	Criterion(a) Name	Severity Rate	Usability Issue	Interdependencies	Recommendations
31. Lack of integration with mobile network	Resource Safety	2	Just as mobile devices are pervasive and essential in users' lives so is their need for constant connection to different networks such as Internet, local or corporate, mobile network, etc. When sending a silent alarm through the MTM's emergency application from their mobile devices, users' devices can be compromised, as can MTM processes.	User satisfaction, performance, and user physical security.	The system's owner should provide as much access as needed to their customers; they should provide connectivity anywhere and at any time, especially in case of emergency. Users can send a Reverse PIN from their mobile devices through Short Message Service (SMS) in case of system failure (i.e. user convenience and physical security). This functionality is provided only to customers who have subscribed to this service in the financial institution.

Notes for Appendix 2

[1] Contactless Technology for Secure Physical Access: Technology and Standards - Key Implementation Considerations, Smart Card Alliance Report, Smart Card Alliance © 2003 https://www.library.ca.gov/crb/rfidap/docs/SCA-Contactless_Technology_Report.pdf

[2] A Central Processing Unit (CPU) is the electronic circuitry within a computer that carries out the instructions of a computer program by performing the basic arithmetic, logical, control and input/output (I/O) operations specified by the instructions.

[3] Magnetic Ink Character Recognition (MICR) is a technology used to verify the legitimacy or originality of paper documents, especially checks. Special ink, which is sensitive to magnetic fields, is used in the printing of certain characters on the original documents.

[4] National Institute of Standards and Technology (NIST), Department of Commerce of United States http://csrc.nist.gov/publications/nistpubs/800-131A/sp800-131A.pdf

[5] Federal Information Processing Standards Publication 197, November 2001. http://csrc.nist.gov/publications/fips/fips197/fips-197.pdf

[6] ADVANCED ENCRYPTION STANDARD (AES) is the successor of DES as standard symmetric encryption algorithm for US federal organizations. AES uses keys of 128, 192, or 256 bits; although, 128 bit keys provide sufficient strength today. It uses 128 bit blocks, and it is efficient in both software and hardware implementations. It was selected through an open competition involving hundreds of cryptographers over several years.

Data Encryption Standard (DES) is the previous "data encryption standard" from the '70s. Its key size is too short for proper security. The 56 effective bits can be brute forced, and that has been done more than ten years ago. DES uses 64-bit blocks, which poses some potential issues when encrypting several gigabytes of data with the same key.

Triple Data Encryption Standard (Triple DES) is a way to reuse DES implementations, by chaining three instances of DES with different keys. 3DES is believed to still be secure because it requires 2^{112} operations which is not achievable with foreseeable technology. 3DES is very slow especially in software implementations because DES was designed for performance in hardware.

[7] Audit log is a chronological sequence of audit records, each of which contains evidence directly pertaining to and resulting from the execution of a business process or system function. Audit records typically result from activities such as transactions or communications by individual people, systems, accounts or other entities.

Appendix 3: Security Severity Ratings and Recommendations for MTM

Problem Description	Criterion(a) Name	Severity Rate	Security Issue	Interdependencies	Recommendations
1. Exposing security holes	Minimal Action	3	Using customization and personalization methods are one of the important methods for obtaining customers' satisfaction. They improve the efficiency and simplicity of usage of the MTMs and enable users to benefit from the machine corresponding to their needs (e.g. service selection options, language, customized dashboard, etc.). However, to achieve the delivery of next-generation services via MTMs such as customization/ personalization, easy integration between end-user and third-party capabilities is needed. Although distributed hardware and software architecture can deliver a fully customizable service, this may open the door for security holes (e.g. JavaScript that runs in the MTM's browser can be exploited to hack accounts of users who visit a particular user profile page).	Third-party integration, and user customization and personalization.	1. Make a risk assessment to what degree of customization/ personalization is realistic at the MTM without compromising security. 2. A hybrid contactless smart card can be used to provide the required level of security for logical access while providing a reliable and easy-to-use solution for physical access. Or use multiple technology cards that can combine either of the ISO/IEC standard contactless smart card technologies with 125 kHz proximity technology. 3. Make a risk assessment to what degree of customization/ personalization is realistic at the MTM without compromising security. 4. A hybrid contactless smart card can be used to provide the required level of security for logical access while providing a reliable and easy-to-use solution for physical access. Or use multiple technology cards that can combine either of the ISO/IEC standard contactless smart card technologies with 125 kHz proximity technology. This enables the card to operate with legacy physical access control systems, as well as new ISO/IEC-compliant systems. Providing multiple read/write capabilities on a card can often assist in providing the tools needed to enable a transition from legacy to new technology applications over time. 5. Organizations should take advantage of the card architecture by linking physical and logical access privileges to increase security within the card.

(Continued)

Problem Description	Criterion(a) Name	Severity Rate	Security Issue	Interdependencies	Recommendations
2. Authentication task overload	Minimal Action	1	Users are required to switch to another authentication method which varies complexity and familiarity with authentication methods artifacts.	Availability of services, efficiency, performance.	Implement single sign-on (SSO) authentication across multiple service providers. This will remove the need for authenticating users at every instance of remote service provision. However, this involves more than the system owner's decisions but also third-party players so that an agreement should be settled in order to provide this feature to users. Any good SSO implementation should have the flexibility to offer the integration of advanced authentication systems in order to boost the security of the user's login. SSO for individual systems tend to store user credentials locally on the workstation. Care must be taken with this approach since administration, as well as usability, issues will arise if this method is used in a network environment. SSO within network environments tend to store credentials on a central server or within a directory in order to provide access to all workstations. Administration and accessibility is thereby maximized. For enhanced flexibility, an SSO application should be able to provide both central server and local methods of credential storage. Additionally, consideration should be given to an implementation that provides integration with biometric devices and other advanced authentication devices to enhance security.
3. Security risks (visual cues in KBA methods)	Minimal Action	3	Attackers have a higher probability of guessing users' passwords, which is called "password guessing attacks" (e.g. exhaustive search or dictionary attacks) if cues are given to users. These attacks are difficult to control and hence pose a major problem in the functioning of KBA-based authentication methods. The likelihood of remembering several passwords is hard. Half the people who say they never write down their passwords usually need to have their passwords reset because they forget the password. Users get frustrated if they cannot login, given that they have a goal to achieve such as cash withdrawal.	Availability of services; user frustration.	Some countermeasures to password-guessing attacks are described below. **Account locking:** After three consecutive unsuccessful login attempts, the account of the particular user is locked for some time. **Delayed Response:** The server provides a delayed response to the user request (e.g. not faster than one answer per second). **Adopt CAPTCHA authentication:** One of its versions allows users to authenticate as a human by recognizing what object is common in a set of images – image-based recognition. This is easy for humans to respond to but rather difficult for computers to answer. It is worth noting that an online attacker is fundamentally a programmed computer.

(Continued)

Problem Description	Criterion(a) Name	Severity Rate	Security Issue	Interdependencies	Recommendations
4. Insecure default cookies	Minimal Action	3	Whenever the system provides fields that already have values, the system should reduce the time users spend typing and improve their accuracy. Users can still override the default values if required.	Performance; availability of services; user satisfaction.	A simple and secure solution still using cookies: the cookie should only be accessible via SSL on a page using the HTTPS protocol. All other default aspects of the cookie remain the same. To set a secure cookie code, use the "secure" option, which creates a secure cookie by setting the "secure" option to true. As mentioned, this solution will only work if the page calling this code uses the HTTPS protocol, otherwise, the cookie will be generated with default options.
5. Unsafe PIN credential	Minimal Action	1	PINs take less time to enter than passwords but are considered low security compared to passwords. Some of the common security threats and risks associated with PINs include the following: **PIN social engineering tactics:** Trick users into handing over their computer-chip card and reveal security details such as the card's PIN. **PIN selection:** Top ten easily guessed MTM PIN codes combination according to DataGenetics[1]: 1234, 1111, 0000, 1212, 7777, 1004, 2000, 4444, 2222, 6969. From four to more than eight PIN codes, there are lots of patterns and repetition. **PIN single-factor authentication:** PINs should not be employed as the only form of authentication for a system access. **PIN change Protocol:** A stolen social security number and other private information (e.g. mother's maiden name), a hacker can call a bank pretending to be a legitimate user requesting a PIN change. Once the attacker has the PIN changed, s/he can access all the user's accounts the system has tied to that PIN number. **Brute-force attacks on PINs:** Hackers employ different methods to disclose users PINs. Computers can be employed to run hundreds of combinations of account numbers and PINs at high speed until the exact PIN is entered.	Availability of services; security policies; productivity.	Common recommendations for fraud prevention include the following: **PIN change protocol:** Anytime a customer calls the customer services to access accounts, information should be requested such as the most recent statement balance or the answer to customer-selected security questions to verify the identity of the customer. Customer representatives should neither provide any personal information nor make changes to a customer's account without verifying their identity nor change customer-selected PINs at any time by telephone. **PIN selection:** The average user must remember a plethora of PINs and passwords. Nevertheless, the best defense against becoming the victim of fraud can be selecting a PIN that is not easily guessed by hackers. **Automated PIN checking:** Lock access to an account after three unsuccessful PIN attempts. To unlock, a customer must call customer services and verify their identity.

(Continued)

Problem Description	Criterion(a) Name	Severity Rate	Security Issue	Interdependencies	Recommendations
6. Diminished security (physical and logical access)	Minimal Action	2	Although a three-factor authentication is used (hand recognition), it is not implemented using a contactless technology, which affects performance and security.	Ubiquitous computing (i.e. technology retreats into the background of our lives), performance, and reliability.	**Physical access control** authenticates users and grants access to physically secure areas while *logical access control* authenticates them and grants access to accounts and networks. For physical access applications, contactless technology provides reliable throughput (i.e. interchange of data). If biometrics is used, the throughput advantages offered by contactless technology are decreased, but the strength of security and authentication is increased. As already stressed, usable security involves trade-offs between security and usability. As the authentication method used is biometric, which is considered quite usable, there's no significant impact from the usability standpoint.

(Continued)

Problem Description	Criterion(a) Name	Severity Rate	Security Issue	Interdependencies	Recommendations
					Applications using contactless smart cards support many security features that ensure the integrity, confidentiality, and privacy of information stored or transmitted[2], including the following: 1. **Enhances Security:** Capability to carry either a digital certificate or a biometric template to enhance authentication of the cardholder's identity. Contactless cards provide the tools to enable more secure access to buildings, secure areas, and electronic systems. They provide secure tokens to hold the key pairs that enable the authentication of the recipient and originator of transactions across public networks, and that can be used to encrypt transactions. In conjunction with contactless smart cards, biometrics can provide strong security for PKI credentials held on the cards, thus providing greater trust in PKI services, especially digital signatures for non-repudiation. 2. **Performance:** With JAVA Card technology and faster processors, the time to read data from and write data to the chip has been greatly reduced. 3. **Mutual authentication:** For applications requiring secure card access, the contactless smart card device can verify that the reader is authentic and can prove its own authenticity to the reader before starting a secure transaction. 4. **Preventing eavesdropping:** Information stored on contactless cards can be encrypted and communication between the card device and the reader can be encrypted to prevent eavesdropping. 5. **Strong contactless device security:** Contactless smart cards have built-in tamper-resistance and are extremely difficult to duplicate or forge.

(Continued)

Problem Description	Criterion(a) Name	Severity Rate	Security Issue	Interdependencies	Recommendations
					7. **Authenticated and authorized information access:** The contactless smart card has the ability to process information and react to its environment allows it to provide authenticated information access and protect the privacy of personal information. The card can verify the authority of the information requestor and then allow access only to the information required. Access to stored information can also be further protected by a personal identification number (PIN) or biometric to protect privacy and counter unauthorized access.
					8. **Strong support for information privacy:** Smart cards ensure the ability of a system to protect individual privacy. Unlike other technologies, smart card-based devices can implement a personal firewall for an individual, releasing only the information required and only when it is required. The ability to support authenticated and authorized information access and the strong contactless device and data security make contactless smart cards excellent guardians of personal information and individual privacy.
7. Absence of remote or local authentication	Operability	2	Information transiting over the Internet (open network) is subjected to attacks. Although the system uses encryption, there is always a security risk. Users can get frustrated if remote or local authentication is not functional. Remember, they have a goal to achieve and they want to do it quickly.	Availability of services; performance.	The system should be set up as local access by default, which represents stronger security. It means that information does not need to be transmitted over the network, and so it avoids the growing threat of online attacks.
8. Insecure "off card" authentication	Operability	2	There is a separate physical access control system database managed by the system owner which maintains an employee's physical access control privileges and issues the proximity card.	Performance; availability of services; efficiency; portability.	Same recommendations as in item 6. Through the use of locking mechanisms and encryption, data stored on smart card chips can be made very secure.

(Continued)

Problem Description	Criterion(a) Name	Severity Rate	Security Issue	Interdependencies	Recommendations
9. Lack of stronger security (authentication factors)	Operability	3	Although a three-factor authentication is used (hand geometry), it is not implemented using a contactless technology, which affects performance and security. The smart card is not used to store the user's biometric data including user profile and enrollment.	Portability; performance; interoperability; flexibility.	Same recommendations as in item 6. The biometric method of match-on-card protects the initial enrollment template since it is maintained within the smart card and never transmitted off-card. A contactless smart card can support biometric authentication. For human identification systems that require the highest degree of security and privacy, a contactless card can be implemented in combination with biometric technology. Smart cards and biometrics are a natural fit to supply two- or multi-factor authentication. A smart card is the logical secure storage medium for biometric information. During the enrollment process, the biometric template can be stored on the smart card chip for later verification. Only the authorized user with a biometric matching the stored enrollment template receives access and privileges.
10. Unsafe PIN length	Security	3	An MTM relies on short, low-entropy PINs for authentication. Entropy is the randomness collected by an operating system or application for use in cryptography or other uses that require random data. Modeling secrecy of PIN is fairly subtle. A PIN is intended to be a human-memorable secret value and therefore has low entropy. In practice, four-digit PINs are used. The main property of low-entropy secrets is that they are guessable. If an attacker is able to guess a PIN value and verify user guess offline, this could constitute a real attack because exhaustively attempting to verify all four-digit numbers is computationally feasible. A four-digit PIN can be broken in less than a second, a six-digit PIN in about 10 seconds, and a ten-digit PIN would likely take weeks to crack.	Performance; efficiency.	ISO 9564-1:2002 allows for PINs from 4 up to 12 digits, but also notes that, for usability reasons, an assigned numeric PIN should not exceed six digits in length. So ideally, use PINs with a large number of digits, for instance a six-digit PIN. A longer PIN obviously provides greater security against an attacker who tries to guess a user's PIN or who tries to read a PIN over the shoulder of a user. Hence, a slightly longer code (more than four digits) is not a hardship given the security benefits.

(Continued)

Problem Description	Criterion(a) Name	Severity Rate	Security Issue	Interdependencies	Recommendations
11. High FAR and FRR for hand geometry recognition	Operability	2	Hand geometry authentication at the MTM currently presents high FRR of 5% and FAR of 3%, which are far from the average rates. Users like hand geometry and fingerprint recognition because they are not intrusive and because they are reliable, low cost, and, in general, only have a 0.1% FRR, which is one of the lowest in the biometrics industry.	Availability of services, performance.	The system should be adjusted to an FRR of about 1.05% and an FAR of about 0.1%. Zhang (2004) has documented the hand geometry recognition system as being able to operate with a low FAR of 0.096% and reasonable FRR of 1.05%. The accuracy of each system can be adjusted, which means that the secret is to balance the likelihoods of FRR and FAR, so the system rarely locks out legitimate users and it doesn't fall for masquerade attacks. However, the main drawback of changing the system sensitivity to reduce FRRs is that it may increase FARs.
12. High error rate of processing errors	Operability	1	In banking, the error rate seems to lie somewhere between 1 in 10,000 and 1 in 100,000 transactions (Anderson, 2008), but this depends on the application domain.	Performance; availability of services.	It's difficult to say whether an exception is due to fraud or to error, and what the error rate range should be. So the lower the transaction error rate, the better. No security policy will ever be entirely rigid; there will always have to be workarounds for people to cope with real life, and some of these workarounds will create vulnerabilities.
13. Disclosure of sensitive business personnel information	Privacy	2	Personnel sensitive information includes personnel, medical, and similar files whose disclosure would constitute a clearly unwarranted invasion of privacy in an organization. Examples include user IDs, passwords, PINs, employee payroll data, tax reports and payments, payments for employee benefit and welfare plans, travel-related costs and information, employee performance information, and medical records.	Disruption of business (i.e. disruption of the MTM owner's ability to generate services and, consequently, revenue); system distrust by employees.	Smart cards can help to protect privacy with secure data storage on the card (e.g. PIN, passwords, and biometric template). These data can only be accessed through the smart card OS by those with appropriate access rights. This functionality can be used by the system to improve privacy (e.g. storing employee data on the card rather than in a central database. Therefore, the employee has better knowledge and control of when and by whom their personal data is being granted access). Given that the majority of the ATMs in the market uses the Triple Data Encryption Standard (3DES) or even the Data Encryption Standard (DES), it is worth noting that 3DES offers a significantly higher security for sensitive information, including PIN mechanism. Also, since 3DES is based on the same algorithm as single DES, it can be implemented into the existing Electronic Funds Transfer (EFT) network with a minimum of disruption. The implementation of 3DES is required with the aim of maintaining users' trust in payment systems and to guarantee the integrity of confidential cardholder information.

(Continued)

Problem Description	Criterion(a) Name	Severity Rate	Security Issue	Interdependencies	Recommendations
14. Compromised private information (different databases from different applications that reside on a multi-application contactless card)	Privacy	2	The system/application owner is the financial institution (e.g. bank) who "owns" the card and "leases" space to the third parties for its users' applications. The financial institution has control over the card specification and operating environment. Also, there is a client registry but the current system does not provide pointers to all application owner databases active on the card. The card contains different applications and different databases coexist in the same environment without any integration. For this reason, if the card is lost, the cardholder will encounter difficulties to recover all applications to repopulate the card.	Enables significant productivity gains; information system liability.	The cardholder ensures the accuracy of personal data; application owners are responsible for protecting personal data provided by the cardholder and maintaining the accuracy of that data. The recommendation is to integrate the multiple databases on the card so that, for example, the content of the badging system, physical access control privilege database, and logical access control privilege database can be amalgamated into a single integrated database maintained as part of the card management system. This approach reduces the need to maintain multiple separate systems and consequently the leakage of sensitive information. In addition, the multi-application card can securely hold multiple application usernames, and passwords, offering the user convenient access through a single PIN (or biometric) and reducing or eliminating the cost of help desk calls. Also, decentralized applications would perform all transactions, but they have shadow files maintained in the centralized database of the application owners. Finally, the system's owner should provide procedures to safeguard the privacy of "shadow" databases, and document these procedures in the card issuer and cardholder agreement.

Problem Description	Criterion(a) Name	Severity Rate	Security Issue	Interdependencies	Recommendations
15. Low security (match off-card technique)	Privacy	2	In a match off-card technique, the enrolled template is originally loaded onto the smart card and then dispensed from the smart card via the contactless interface when requested by the external biometric system. The external equipment then compares a new live scan template of the biometric with the one being presented from the smart card. This implementation obviously has some security risks related to transmitting the enrolled template off the smart card for every biometric challenge. The system's owner should ensure the confidentiality and integrity of the released template. With the match off-card technique, the smart card is storing a template (or multiple templates) but has no major knowledge of the kind of biometric information, nor the capability to process it in any manner.	Interoperability; performance.	The recommendation is to use the match on-card technique. Match-on-card technique originally stores the enrollment template into the smart card's secure memory. When a biometric match is requested, the external equipment submits a new live scan template to the smart card which then carries out the matching operation within its secure processor and securely communicates the outcome to the external equipment (i.e. when the smart card itself is employed to carry out the one-to-one identity verification rather than external equipment, a high degree of confidence and security of the credential's verification is achieved). This method protects the original enrollment template since it is maintained within the smart card and never transmitted off-card. Matching hand geometry information in the card removes the uncertainty of matching on a network-connected device, an external server, or a database. These could be regarded as weak links in a security chain. User privacy is also maintained with this technique since the user's biometric template information is not readable from the card. With this technique, the card must be a microcontroller-based device and be capable of computing the one-to-one match.

(Continued)

Problem Description	Criterion(a) Name	Severity Rate	Security Issue	Interdependencies	Recommendations
16. Compromised security (security and financial applications in single functions application cards)	Security	2	A single card used for different purposes runs the risk of creating a centralized storehouse of data about users' activities (e.g. banking, medical, credit cards transactions records, etc.).	Software liability; scalability; system trust.	Contactless multi-application smart card technology is used in applications that require protecting personal information and/or delivering fast, secure transactions. The majority of manufacturers write proprietary OSs for each category of chip card they produce. However, there are also multi-application OSs which are placed on top of the card's proprietary system and can host parallel applications separated by secure firewalls. The Java Card platform provides a secure execution environment with a firewall between different applications in the same card (e.g. IT security: logon to networks, digital signature, biometrics, and encryption, banking and finance, government, corporation applications, etc.). This allows diverse applications on the same card to function independently from each other as if they were on separate cards. The "walls" between the individual's activities records protect individual privacy in two ways: they restrict the damage to individual privacy that takes place through either misuse by an authorized user or unauthorized access by an attacker, and they place auditing on the monitoring capacity of each system. Security features that ensure the integrity, confidentiality, and privacy of information stored or transmitted include smart cards which provide secure communications between the card and card readers; this feature allows smart cards to send and receive data in a secure and private manner, and it can be used by a system to enhance privacy by ensuring that data sent to and from the card is not intercepted or tapped into. With regard to software liability, best practices recommend that software vendors protect themselves from legal liability for defective software by investing in improved product quality and by relying on legal protections – the authors consider that these two strategies are not mutually exclusive.

(Continued)

Problem Description	Criterion(a) Name	Severity Rate	Security Issue	Interdependencies	Recommendations
17. Lower security (inappropriate security algorithms)	Security	4	Currently, the system does not employ the algorithms specified within the (FIPS 140-2_02) FIPS 140-2: Security Requirements for Cryptographic Modules.	System trust; scalability; performance.	Meet the security requirements indicated in (FIPS 140-2_02) FIPS 140-2: Security Requirements for Cryptographic Modules.[3]
18. Missing additional authentication method (KBA over the telephone)	Security	3	The risk of eavesdropping through a telephone network is a genuine threat, particularly since it cannot be encrypted.	Flexibility; availability of services; user satisfaction.	Introduce an additional authentication method to be used in conjunction with a PIN, like voice recognition. The ideal recommendation is to switch to Voice over Internet protocol (VoIP) phone calls. It means that the system would convert a voice signal into data packets and send them over the Internet. VoIP converts conversations to packets of bits that can be effortlessly encrypted with secret keys.
19. Compromised response time and performance (system hardware component)	Security	2	Processor currently used is not adequate to the high demands of smart card technology.	Availability of services; task abandonment.	An example of a faster processor is the ARM® Cortex®-R8[4] processor; it is the highest performance Cortex-R series processor, delivering advanced real-time processing for a wide range of demanding and deeply embedded applications. It provides fast real-time handling of multiple interfaces for new high-speed and contactless applications. Another solution is to employ the Java Card Technology.

(Continued)

Problem Description	Criterion(a) Name	Severity Rate	Security Issue	Interdependencies	Recommendations
20. Security risks (WMLScript language)	Security	2	As of this writing (February 2016), WML is becoming less and less common. This technology is still used in developing countries in older mobile phones; also, coding up modern pages in WML will increase access in the developing world. Possible attacks are: i) Theft or damage of personal information; ii) Abusing user's authentication information; iii) Malicious scripts will have the ability to falsely ring up charges or potentially off-load money from smartcards or bank accounts. Some of the security risks of WMLScripts language: • WMLScripts is not a type-safe language. Type safety is a property of some programming languages that involve the use of a type system to prevent certain erroneous or undesirable program behavior called type errors. • WMLScripts can be scheduled to be pushed to the client device without the user's knowledge. • WMLScripts language doesn't prevent access to persistent storage.	Availability of services; applications liability; system trust.	The MTM hosts different third-party applications so that users can access some of them over the Internet through the MTM. However, the security of every single third-party application cannot be guaranteed by the system's owner. The recommendation is that the system's owner should evaluate third-parties applications and its security risks before implementing them on the MTM. Some of the best practices are: • MTMs should not allow unauthorized applications to communicate. • MTMs should verify application integrity. • MTM should be invisible to viruses, worms, and hackers. • Lock-down the MTM allowing only authenticated applications and data transfer to execute between trusted endpoints.

(Continued)

Problem Description	Criterion(a) Name	Severity Rate	Security Issue	Interdependencies	Recommendations
21. Lack of response time (smart card)	Load Time	2	A smart card with a slow response time can lead to a substantial performance decrease. For example, consider this scenario: the threat from an untrusted card reader that the user sticks her card into. This cardholder will not sit patiently while the reader seems not to be working. If she inserts her smart card into a reader and nothing happens for more than a few seconds (e.g. 2 or 3 seconds), she will likely pull the card out and try again. If nothing happens yet again, she will find another reader. The slow response time the card merely allows a fraudulent reader does so much damage before the cardholder removes her card.	Availability of services; task abandonment.	A high-performance FIPS[5] certified smart card with a separate processor and cryptographic chip and memory for encryption as well as a 32-bit CPU is the recommended solution for this issue.
22. Lack of system integration (check imaging)	Load Time	2	Without ensuring auditing and data integrity capabilities to the system, users and the system's owner will suffer from potential attacks to the check imaging process.	Availability of services; task abandonment.	The system should integrate the check imaging workflow process to the IT infrastructure to easily detect fraud and facilitate auditing and reporting.

(Continued)

Problem Description	Criterion(a) Name	Severity Rate	Security Issue	Interdependencies	Recommendations
23. Lack of security (check imaging)	Load Time	3	With various forms of check fraud already on the rise, the already common check imaging stands to have an even greater impact on the security of the system's owner. Though check imaging can certainly contribute to cost savings and increased convenience for customers and the system's owner alike, it introduces major risks to everyone involved since these digital checks include all of the printed version's sensitive information, with little of its security features.	Availability of services; system trust; performance; flexibility.	**Make Alterations Detectable:** System's owner should offer an image security feature which makes image alterations detectable by creating an exclusive mathematical digital signature on the original image. As a result, in the image exchange process, the system's owner can recognize even the slightest alteration to an image. **Digital signature:** The system's owner should check processing hardware and image security software to generate an exclusive digital identity for each check image created. The mathematical digital signature is attached to the image file at the point of capture and stays with the image as it moves through the payment system. Benefits of image security: detects fraud quickly, reduces the risks associated by image exchange, generates a secure identity for each check image, and generates an unquestionable link between the image and the imaging system's owner.
24. Lower security (DES encryption algorithm)	Load Time	3	Same as item 27. User privacy and Integration – II. Usability Severity Ratings and Recommendations for MTM.	Availability of services; performance.	Same as item 27. User privacy and Integration–II. Usability Severity Ratings and Recommendations for MTM.

(Continued)

Problem Description	Criterion(a) Name	Severity Rate	Security Issue	Interdependencies	Recommendations
25. Weak auditing capabilities	Load Time	2	Log management deals with large volumes of computer-generated log messages (also known as audit records, audit trails, event logs, etc.). Concerns about security, system and network operations (i.e. system or network administration) and regulatory compliance drive log management. Log management covers: log collection, centralized aggregation, long-term retention, log rotation, log analysis (in real time and in bulk after storage), log search, and reporting. Auditing user activity provides the auditor with the assurance that the policies, procedures, and safeguards that management has established are working as intended. However, the system does not provide customizable auditing and reporting capabilities only automatic logs, which also includes accidental actions.	Availability of services; system trust; performance.	As already mentioned, the system should integrate the check imaging workflow process to the IT infrastructure to easily detect fraud and facilitate auditing and reporting. One of the categories of internal audit review falls under the Information Systems (IS) Audits which address the internal control environment of automated information processing systems and how these systems are used. IS audits typically evaluate system input, output and processing controls, backup and recovery plans, and system security, as well as computer facility reviews. The most important audit files that should be analyzed online is file server's security audit log files. Some of the main vendors in log management are: LogRhythm, McAfee LogMatrix, IBM Security QRadar Log Manager, Splunk, etc.
26. Lower PIN security	Minimal Memory Load	2	A PIN should be a number that users will easily remember. However, it also should not be a number that others can easily guess. Users should avoid their birth date, social security number, telephone number, street address, etc. Because PINs are used often, users may not use consecutive numbers such as 1234 or recurring numbers such as 1111.	Availability of services; performance on user authentication.	A fast additional authenticator is the use of "security questions." Security questions are commonly used as authenticator by banks, institutions, and organizations as an additional security layer. They are a type of shared secret. Users sometimes need to remember the precise spelling and cases of answers they provide to the security questions. The best answers are straightforward, memorable, can't be guessed easily, and don't modify over time. For example, a financial institution could ask for a customer: "Where does your nearest sibling live?" before issuing a replacement for a lost debit card. Adopting security questions in conjunction with PIN represents a strong authentication (two-factor authentication).

(Continued)

Problem Description	Criterion(a) Name	Severity Rate	Security Issue	Interdependencies	Recommendations
27. Lack of enforcement of security policy	Resource Safety	4	MTMs offer a real convenience, but at the same time, they offer an element of risk. Many of the MTM robberies occur after the cash withdrawal. Therefore, the system's owner should provide a resource safety countermeasure for its customers and enforce its security policy. Without security policies, no enforcement of security standards or configurations is able to be made. By establishing a policy, it is supposed that enforcement can or will pursue.	Availability of services; system trust; user confidence and satisfaction.	**MTM Reverse PIN:** After the user enters the correct PIN in reverse order, the account holder's account is automatically locked and the system triggers an error message on the screen saying that the user's account has been locked. At the same time, the system also dials the preconfigured National Security Force number to have the robber trapped. This is clearly a resource safety feature that an MTM offers to its customers. To unlock user's account, he will be required to make a call to customer service and provide, for example, security questions as an additional authentication mechanism and/or confirm his private information stored in the system owner's database.
28. Lack of wireless attacks prevention	Resource Safety	2	As wireless devices are pervasive and essential, they are both a target for attack and also a weapon with which such an attack can be performed. So it is the responsibility of the system's owner to offer security in all points of interaction between the MTM and its components, such as the silent alarm application installed on the user's mobile device. When sending a silent alarm from an MTM emergency app, users' devices can be compromised as well as MTM processes.	Availability of services; system trust; physical security.	A super administrator is able to wipe information within the wireless device or disable a lost or stolen one with Over-The-Air (OTA) commands. Remote device wipe initiates and tracks a remote wipe command for lost or stolen mobile devices. Remote device wipe is a feature that enables a server to set a mobile device to delete all data the next time that device connects to the server. OTA is a standard for the transmission and reception of application-related information in a wireless communications system. OTA is commonly used in conjunction with the Short Messaging Service (SMS), which allows the transfer of small text files even while using a mobile phone for more conventional purposes. In addition to short messages and small graphics, such files can contain instructions for subscription activation, banking transactions, ringtones, and Wireless Access Protocol (WAP) settings. OTA messages can be encrypted to ensure user privacy and data security.

[1] PIN Analysis. DataGenetics blog, February 13, 2016, http://www.datagenetics.com/blog/september32012

[2] Smart Cards Alliance. Smart Cards Talk, Vol. 9, Nr. 12, December 2004 http://www.smartcardalliance.org/newsletter/december_04/feature_1204.html

[3] Standards - FIPS PUB 140-2 - Effective 15-Nov-2001 Security Requirements for Cryptographic Modules: http://csrc.nist.gov/groups/STM/cmvp/standards.html#02 Implementation Guidance for FIPS PUB 140-2 and the Cryptographic Module Validation Program: <http://csrc.nist.gov/groups/STM/cmvp/documents/fips140-2/FIPS1402IG.pdf> Links retrieved on February 21, 2016.

[4] ARM Inc. Retrieved on February 21, 2016, http://www.arm.com/products/processors/cortex-r/cortex-r8-processor.php

[5] Standards - FIPS PUB 140-2 - Effective 15-Nov-2001 Security Requirements for Cryptographic Modules: <http//csrc.nist.gov/groups/STM/cmvp/standards.html#02> Implementation Guidance for FIPS PUB 140-2 and the Cryptographic Module Validation Program: http://csrc.nist.gov/groups/STM/cmvp/documents/fips140-2/FIPS1402IG.pdf Links retrieved on February 21, 2016.

Additional Reading

Alfonzo, O., Domínguez, K., et al. 2008. Quality Measurement Model for Analysis and Design Tools based on FLOSS. *19th Australian Conference on Software Engineering*, 1530-0803/08, IEEE. DOI 10.1109/ASWEC.2008.70.

Allan, A. 2007. "WWW.Authentication: Why? When? What? Who?" Gartner Identity Access Management Summit, LA. Gartner, Track Session. http://agendabuilder. gartner.com/iam2/webpages/SessionDetail.aspx?EventSessionId=811. Retrieved on February 2, 2007.

Angeli, A. D., Coventry, L., Johnson, G. and Coutts, M. 2003. "Usability and User Authentication: Pictorial Passwords vs. PIN". *Contemporary Ergonomics*, pp. 253–258. London, England: Taylor and Francis.

ANSI. 1998. "Triple Data Encryption Algorithm Modes of Operation". ANSI X9.52-1998. Committee X9 (Financial Services). Online. http://webstore.ansi.org/ RecordDetail. aspx?sku=ANSI+X9.52%3A1998. Retrieved on January 5, 2007.

Baddeley, A. D. 1998. *Human Memory: Theory and Practice*. Boston, MA: Allyn and Bacon.

Bastien, J. M. C. and Scapin, D. L. 1993. "Ergonomic Criteria for the Evaluation of Human-Computer Interfaces". *Behaviour and Information Technology*, 16(4–5), pp. 220–231.

Biederman, I. 1987. "Recognition-by-Components: A Theory of Human Image Understanding". *Psychological Review Journal*, 94(2), pp. 115–47.

Calvary, G., Coutaz, J., Bouillon, L., Florins, M., Limbourg, Q., Marucci, L., Paternò, F., Santoro, C., Souchon, N., Thevenin, D., Vanderdonckt, J., The CAMELEON Reference Framework, Deliverable 1.1, CAMELEON Project.

Common Vulnerabilities and Exposures (CVE®). 2009. "Common Vulnerabilities and Exposures". http://cve.mitre.org/index.html. Retrieved on March 6, 2009.

Dix, A., Finlay, J., et al. 1993. *Human-Computer Interaction*, Englewood Cliffs: Prentice Hall.

Jeffery K. &, Neidecker-Lutz B. 2009. The Future of Cloud Computing –. Opportunities for European Cloud Computing beyond 2010. https://ec.europa.eu/digital-single-market/en/news/future-cloud-computing-opportunities-european-cloud-computing-beyond-2010-expert-group-report

Jøsang, A. 2006. "Trust and Reputation Systems, In Alessandro Aldini", Roberto Gorrieri (Eds.): *Foundations of Security Analysis and Design IV, FOSAD* 2006/2007 pp. 209–245.

Klein, G. A. 1998. *Sources of Power: How People Make Decisions*. Cambridge, Massachusetts: MIT Press.

Landgren, J. 2007. Designing information technology for emergency response. Gothenburg Studies of Informatics, Report 39.

Mendoza, L. E., Perez, M. A., et al. 2005. " Prototipo de un modelo sistémico de calidad (MOSCA) del software." *Computación y Sistemas*, 8(3), pp. 196–221.

Paternò F. 2000. *Model-Based Design and Evaluation of Interactive Applications*, Springer Verlag.

Paternò F., Santoro C., Spano L.D. 2003. MARIA: A Universal Declarative Language for Service-Oriented Applications in Ubiquitous Environments, *ACM Transactions on Computer-Human Interaction*, November 2009, pp. 19:1–19:30, ACM Press.

Preece, J., Rogers, Y., et al. 1994. *Human Computer Interaction*. Wokingham, UK: Addison-Wesley.

Sasse, M. A, Flechais, I. and Mascolo, C. 2007 "Integrating Security and Usability into the Requirements and Design Process". *International Journal of Electronic Security and Digital Forensics Archive*, 1(1).

Schultz, E. 2007. Research on Usability in Information Security, *Computer Fraud and Security*, 2007 (6), pp. 8–10, doi:10.1016/S1361-3723(07)70075-1.

Shneiderman, B. 1998. *Designing the user Interface: Strategies for Effective Humancomputer*, Reading, MA: Addison Wesley Longman.

Turoff, M., Chumer, M., Van de Walle, B., Yao, X. 2004. The Design of a Dynamic Emergency Response Management Information System (DERMIS). *Journal of Information Technology Theory and Application (JITTA)*, 5 (4), pp. 1–36.

Vargas, L. S., Gutiérrez, A. F., et al. 2008. MECRAD: Model and Tool for the Technical Quality Evaluation of Software Products in Visual Environment. *The Third International Multi-Conference on Computing in the Global Information Technology*, 978-0-7695-3275-2/08, IEEE, doi: 10.1109/ICCGI.2008.50.

Winter, S., Wagner, S., et al. 2008. A Comprehensive Model of Usability, *LNCS*, 4940, pp. 106–122.

References

Abran, A., Khelifi, A., Suryn, W. and Seffah, A. 2003. "Usability Meanings and Interpretations in ISO Standards". *Software Quality Journal*, 11(4), pp. 325–338.

Accenture Consulting. 2004. "Guiding Principles – Security Framework". http://www.biz-tech.pl/wbi/Anders_Carlstedt.pdf. Retrieved on March 23, 2006.

Adams, A. and Sasse, M. 1999. "Users Are Not the Enemy". *Communications of the ACM*, 42(12), pp. 40–46.

Ahn, L. V., Blum, M., Hopper, N. J. and Langford, J. 2003. "CAPTCHA: Telling Humans and Computers Apart". *Advances in Cryptology, Eurocrypt'03, Lecture Notes in Computer Science*, 2656, pp. 294–311.

Alexander, C., 1979. "*The Timeless Way of Building*", New York: Oxford University Press.

Alexander, C., Ishikawa, S., and Silverstein, M., 1977. "*A Pattern Language: Towns, Buildings, and Constructions.*" New York: Oxford University Press.

Allan, A. 2007. "WWW.Authentication: Why? When? What? Who?" *Gartner Identity Access Management Summit*, LA. Gartner, Track Session. http://agendabuilder.gartner.com/iam2/webpages/SessionDetail.aspx?EventSessionId=811. Retrieved on February 2, 2007.

Allan, A., Singh, A. and Ahlm, E. 2014. "Magic Quadrant for User Authentication", published 1 December 2014 - ID G00260746.

Alshammari, B., Fidge, C., et al. 2009. "Security Metrics for Object-Oriented Class Designs." *2009 Ninth International Conference on Quality Software*, pp. 11–20.

Anderson, R. C. 1977. "The Notion of Schemata and the Educational Enterprise: Discussion of the Conference". In R. C. Anderson, R. J. Spiro, and W. E. Montague (Eds.) *Schooling and the Acquisition of Knowledge*. Hillsdale, NJ: Lawrence Erlbaum Associates. pp. 415–431.

Anderson, R. J. 2008. *Security Engineering: A Guide to Building Dependable Distributed Systems, Second Edition*. Indianapolis, IN: John Wiley & Sons, p. 1040 .

Angeli, A. D., Coventry, L., Johnson, G. and Coutts, M. 2003. "Usability and User Authentication: Pictorial Passwords vs. PIN". *Contemporary Ergonomics*, pp. 253–258. London, England: Taylor and Francis.

Apple Inc. 2008. "Apple Human Interface Guidelines". http://developer.apple.com/documentation/UserExperience/Conceptual/AppleHIGuidelines/OSXHIGuidelines.pdf. Retrieved on October 3, 2009.

Atkinson, R. C. and Shiffrin, R. M. 1968. "Human Memory: A Proposed System and its Control Processes". In K. W. Spence and J. T. Spence (Eds.), *The Psychology of Learning and Motivation: Advances in Research and Theory*, 2, pp. 89–195.

Baddeley, A. D. 1998. *Human Memory: Theory and Practice*. Boston, MA: Allyn and Bacon.

Balfanz, D., Smetters, D. K. and Grinter, R. E. 2004. "In Search of Usable Security: Five Lessons from the Field". *IEEE Security and Privacy*, 2(5), pp. 19–24.

Bank of America Corporation. 2009. "SiteKey® at Bank of America". http://www.bankofamerica.com/privacy/sitekey. Retrieved on April 13, 2009.

Bastien, J. M. C and Scapin, D. L. 1993. "Ergonomic Criteria for the Evaluation of Human-Computer Interfaces". *Behaviour & Information Technology*, 16(4–5), pp. 220–231.

Bevan, N. 1995. "Measuring Usability as Quality of Use." *Software Quality Journal*, 4, pp. 115–150.

Bevan, N. 2010. Los nuevos modelos de ISO para la calidad y la calidad en uso del software. Calidad del producto y proceso software. C. Calero, M. Á. Moraga and M. Piattini, *RAMA:* pp. 11–34.

Bias, R. G., Mayhew, D. J. 2005. Cost-Justifying Usability, *An Update for the Internet Age, Second Edition (Interactive Technologies)*, Morgan Kaufmann Publishers.

Biederman, I. 1987. "Recognition-by-Components: A Theory of Human Image Understanding". *Psychological Review Journal*, 94(2), pp. 115–47.

Bolle, R., Connell, J., Pankanti, S., Ratha, N. and Senior, A. 2004. *Guide to Biometrics*. New York, NY: Springer-Verlag.

Bovair, S., Kieras, D. E., and Polson, P. G. 1990. "The Acquisition and Performance of Text Editing Skill: A Cognitive Complexity Analysis". *Human-Computer Interaction*, 5(1), p. 48.

BPEL–Web Services Business Process Execution Language, Version 2.0. 2007. OASIS http://docs.oasis-open.org/wsbpel/2.0/wsbpel-v2.0.html.

Braz, C. and Aïmeur, E. 2003. "AuthenLink: A User-Centred Authentication System for a Secure Mobile Commerce". Master Thesis, University of Montreal, Montreal, QC. p. 101.

Braz, C. and Aïmeur, E. 2004. "AuthenLink: Authentication System for a Secure Mobile". *3rd International Workshop on Wireless Information Systems (WIS-2004)*, Porto, Portugal. pp. 114–126.

Braz, C. and Aïmeur, E. 2005. "ASEMC: Authentication for a Secure Mobile Commerce". *RFID Journal*, White Papers, Security.

Braz, C. and Robert, J. M. 2006. "Security and Usability: The Case of User Authentication Methods". *18th French-Speaking Conference on Human Computer Interaction (HCI2006)*, Montreal, Canada. pp. 199–203.

Braz, C., Poirier, P. and Seffah, A. 2010. "Integrating a Usable Security Protocol for User Authentication into the Requirements and Design Process". Doctoral Thesis, University of Quebec at Montreal, QC, Canada. p. 498 p.

Braz, C., Seffah, A. and M'Raihi, D. 2007. "Designing a Trade-off Between Usability and Security: A Metrics Based-Model". *Interact 2007: Socially Responsible Interaction, IFIP TC.13 IFIP Technical Committee on Human Computer Interaction*, Rio de Janeiro, Brazil. pp. 114–126.

Braz, C. and Seffah, A. 2016. "GlanceID: A Usable and Strong Optical User Authentication Method". In 7th International Conference on Applied Human Factors and Ergonomics (AHFE) 2016, Walt Disney World Swan and Dolphin Hotel, Florida, USA July 27–31, 2016.

Buyya, R., Yeo, C. S., Venugopal, S., Broberg, J., Brandic, I. 2009. Cloud Computing and Emerging IT Platforms: Vision, Hype, and Reality for Delivering Computing as the 5th Utility. *Future Generation Computer Systems*, 25(6), pp. 599–616.

Card, S., Moran, T. P. and Newell, A. 1980. "The Keystroke-level Model for User Performance Time with Interactive Systems". *Communications of the ACM*, 23(7), pp. 396–410.

Card, S., Moran, T. P., and Newell, A. 1983. *The Psychology of Human-Computer Interaction*. Hillsdale, NJ: Lawrence Erlbaum Associates.

Castano, S., Fugini, M., Martella, G. and Samarati, P. 1994. A new approach to security system development. Castano, S., Martella, G., Samarati, P. In Proceeding NSPW '94 Proceedings of the 1994 workshop on New security paradigms, pp. 82–88.

Castano, S., Fugini, M. G., Martella, G., Samarati. P. 1995. Database Security. Addison-Wesley & ACM Press.

Cavoukian, A. 2008. Privacy in the clouds, Identity in the Information Society, Volume 1, Number 1, doi: 10.1007/s12394-008-0005-z, http://www.springerlink.com/content/e13m644537204002/fulltext.pdf.

CCN-CERT 2008. Últimos avances en ciberseguridad (9th NATO cyberdefense workshop). Revista auditoria y Seguridad (www.revista-ays.com). June 23:70–71.

Chandra, S., Khan, R. A., et al. 2009. "Security Estimation Framework: Design Phase Perspective." *2009 Sixth International Conference on Information Technology - New Generations*: pp. 254–259.

Chang, E., Dillon T., Calder, D. 2008. Human system interaction with confident computing - The mega trend, *Conference on Human System Interactions*, pp. 1–11, http://ieeexplore.ieee.org/stamp/stamp.jsp?tp=&arnumber=4581399&isnumber=4581395.

Chiasson, S. and Biddle, R. 2007. "Issues in User Authentication". *CHI Workshop: Security User Studies: Methodologies and Best Practices, ACM CHI 2007*, San Jose, CA.

Chiasson, S., Biddle, R. and Somayaji, A. 2007. "Even Experts Deserve Usable Security: Design Guidelines for Security Management Systems". *Workshop on Usable IT Security Management (USM'07)*, Pittsburgh, PA.

Chiasson, S., van Oorschot, P. C. and Biddle, R. 2006. "A Usability Study and Critique of Two Password Managers". *15th USENIX Security Symposium*, Vancouver, Canada. pp. 1–16.

Chipman, S. F., Schraagen, J. M. and Shalin, V. L. 2000. Introduction to cognitive task analysis. In J. M Schraagen, S. F. Chipman & V. J. Shute (Eds.), Cognitive Task Analysis (pp. 3–23). Mahwah, NJ: Lawrence Erlbaum Associates.

Chowdhury, I. and Zulkernine, M. 2010. "Using Complexity, Coupling, and Cohesion Metrics as Early Indicators of Vulnerabilities." *Journal of Systems Architecture*.

Christey, S. 2007. "Unforgivable Vulnerabilities. Common Vulnerabilities and Exposures, Documents". The MITRE Corporation. http://cve.mitre.org/docs/docs-2007/unforgivable.pdf. Retrieved on May 25, 2008.

CIO.com. 2009. "12 Top Popular Applications with Critical Security Vulnerabilities". http://www.cio.com/article/477470/_Top_Popular_Applications_with_Critical_Security_Vulnerabilities. Retrieved on June 9, 2009.

Computer Security Institute (CSI). 2008. "CSI – Computer Crime and Security Survey". http://i.cmpnet.com/v2.gocsi.com/pdf/CSIsurvey2008.pdf. Retrieved on July 5, 2009.

Computing Technology Industry Association (CompTIA). 2002. "Committing to Security: A CompTIA Analysis of IT Security and the Workforce". Security survey.

Coram, T. and Lee, J 1999. Experiences: A Pattern Language for User Interface Design, [Online] available at: http://www.maplefish.com/todd/papers/experiences.

Corner, M. D. and Noble, B. D. 2002. "Zero-Interaction Authentication". *MOBICOM'02*, Atlanta, GA. pp. 1–11.

Constantine, L. L. and Lockwood, A. D. L. 1999. Software for Use: A Practical Guide to the Models and Methods of Usage-Centered Design (paperback) 1st Edition, Addison-Wesley Professional (April 17, 1999).

Cranor, L. F. and Garfinkel, S. L. 2005. *Security and Usability: Designing Secure Systems that People Can Use*. Sebastopol, CA: O'Reilly Media Inc.

Dallas Semiconductor. 2006. "iButton: Touch the Future". http://www.maxim-ic.com/products/ibutton/. Retrieved on August 12, 2008.

Desurvire, H. W. 1994. "Are Usability Inspection Methods as Effective as Empirical Testing?" In Nielsen, J., and Mack, R. L. (Eds.), *Usability Inspection Methods*. New York, NY: John Wiley & Sons.

Desurvire, H. W., Kondziela, J. M. and Atwood, M. E. 1992. "What is Gained and Lost When Using Evaluation Methods Other than Empirical Testing". In Monk, A., Diaper, D., and Harrison, M.D. (Eds.), *People and Computers*, 7, pp. 89–102.

Dhamija, R. and Perrig, A. 2000. "Déjà Vu: A User Study – Using Images for Authentication". *Proceedings of the 9th USENIX Security Symposium*, Denver Colorado. 4(4).

Dhamija, R. and Tygar, J. D. 2005. "The Battle Against Phishing: Dynamic Security Skins". *SOUPS'05: Proceedings of the 2005 symposium on Usable privacy and security*, Pittsburg, PA. pp. 77–78.

Dhamija, R., Tygar, J. D. and Hearst, M. 2006. "Why Phishing Works". *CHI '06: Proceedings of the SIGCHI conference on Human Factors in computing systems*, Montreal, Canada. ACM Special Interest Group on Computer-Human Interaction. pp. 581–590.

Diaper, D. and Stanton, N. 2003. *The Handbook of Task Analysis for Human-Computer Interaction*. Mahwah, NJ: Lawrence Erlbaum Associates.

Diffie, W. and Hellman, M. E. 1976. "New Directions in Cryptography". *IEEE Transactions on Information Theory*. 22, pp. 644–654.

Donahue, G. M. 2001. Usability and the Bottom Line. *IEEE Software*, 18, (1): pp. 31–37.

Dowell, J. and Long, J. 1998. "Conception of the Cognitive Engineering Design Problem". *Ergonomics*, 41(2), pp. 126–139.

Duyne D. K. Van, Landay, J. A, and Hong J. I. 2003. *"The Design of Sites: Patterns, Principles, and Processes for Crafting a Customer-Centered Web Experience"*, Addison-Wesley.

El Bekai A., Rossiter N. 2005. A Tree Based Algebra Framework for XML Data Systems, *ICEIS 2005, Proceedings of the Seventh International Conference on Enterprise Information Systems*, Miami, USA, May 25–28, 2005.

EMC Corporation. 2006. EMC VPN Client (Version 4.8.02.0010) Software. Hopkinton, MA: EMC Corp.

EMC Corporation. 2009. "Password Guidelines, Information Security Awareness, Supporting Materials".

Engelberg, D. and Seffah, A. 2002. Design Patterns for the Navigation of Large Information Architectures, *11th Annual Usability Professional Association Conference*, Orlando, Florida, July 8–12, 2002.

Federal Aviation Administration (FAA). 1998. "Report of the Computer-Human Interface Re-Evaluation of the Standard Terminal Automation Replacement System Monitor and Control Workstation". Human Factors Team. http://www.hf.faa.gov/docs/508/docs/STARS-mc.pdf. Retrieved on March 29, 2008.

Federal Financial Institutions Examination Council (FFIEC). 2005. "Interagency Guidance on Authentication in an Internet Banking Environment". http://www.ffiec.gov/ffiecinfobase/resources/info_sec/2006/frb-sr-05-19.pdf. Retrieved on March 4, 2006.

FIPS. 2002. "Security Requirements for Cryptographic Modules" FIPS PUB 140-2. Online Standard. http://csrc.nist.gov/publications/fips/fips140-2/fips1402.pdf. Retrieved on December 8, 2006.

Flechais, I., Sasse, A. M. and Hailes, S. M. V. 2003. "Bringing Security Home: A Process for Developing Secure and Usable Systems". *Workshop on New Security Paradigms*, pp. 49–57. Ascona, Switzerland: ACM Press.

Folmer, E. and Bosch, J. 2003. Usability Patterns in Software Architecture. *Proceedings of the 10th International Conference on Human-Computer Interaction*. Volume I.

Forrester Research, Inc. 2007. "Enterprise and SMB Security Survey, North America and Europe Q3 2007". http://www.forrester.com/ER/Research/Survey/Excerpt/0,, 641,00.html Retrieved on June 2, 2007.

Forsberg, M. 2003. "Why is Speech Recognition Difficult?". Chalmers University of Technology.

Furnell, S., Papadopoulos, I. and Dowland, P. 2003. "A Long-Term Trial of Alternative User Authentication Technologies". *Information Management & Computer Security*, Emerald Group Publishing Limited, 12(2), pp. 178–190.

Gad, R., El-Fishawy, N., El-Sayed, A. and Zorkany, M. 2015. "Multi-Biometric Systems: A State of the Art Survey and Research Directions". In (IJACSA) International Journal of Advanced Computer Science and Applications Vol. 6, No. 6, 2015.

Gamma, E., Helm, R., Johnson, R. and Vlissides, J. 1995. *"Design Patterns: Elements of Reusable Object-Oriented Software"*. Reading, MA: Addison-Wesley.

Garfinkel, S. L. 2005. "Design Principles and Patterns for Computer Systems That Are Simultaneously Secure and Usable". Doctoral Thesis, Massachusetts Institute of Technology, Cambridge, MA, p. 472.

Gautam, N., Chinnam, R.B. and Singh, N. 2007. "Design Reuse Framework: a Perspective for Lean Development". *International Journal of Product Development*, 4(5), pp. 485–507.

Gong, R. J. 1993. "Validating and Refining the GOMS Model Methodology for Software User Interface Design and Evaluation." Doctoral Thesis, University of Michigan.

Gong, R. and Elkerton, J. 1990. "Designing Minimal Documentation Using a GOMS Model: A Usability Evaluation of an Engineering Approach". *Proceedings of the SIGCHI conference on Human factors in computing systems: Empowering people*, Seattle, WA. pp. 99–107, New York, NY: ACM Press.

Gong, R. and Kieras, D. 1994. "A Validation of the GOMS Model Methodology in the Development of a Specialized, Commercial Software Application". In B. Adelson, S. Dumais & J. Olson (Eds.), ACM *CHI'94 Conference on Human Factors in Computing Systems*, Boston, MA. 1, New York, NY: ACM Press. pp. 351–357.

Granlund, A. and Lafrenière, D. 1999. "A Pattern-Supported Approach to the User Interface Design Process", *Workshop Report, UPA'99 Usability Professionals' Association Conference*, Scottsdale, AZ, 29 June–2 July 1999.

Gray, W. D., John, B. E. and Atwood, M. E. 1993. "Project Ernestine: A Validation of GOMS for Prediction and Explanation of Real-world Task Performance". *Human-Computer Interaction*, 8(3), pp. 237–209.

Grudin, J. 1989. "The Case Against User Interface Consistency". *Communications of the ACM*, 32(10), pp. 1164–1173.

Hackos, J. T. and Redish, J. C. 1998. *User and Task Analysis for Interface Design*. New York, NY: John Wiley & Sons.

Harris, D. 2007. "Engineering Psychology and Cognitive Ergonomics". *7th International Conference on Engineering Psychology and Cognitive Ergonomics (EPCE 2007)* in the framework of the *12th International Conference on Human-Computer Interaction (HCI 2007)*, Beijing, China. 4562, Berlin, Germany: Springer.

Hassenzahl, M., Tractinsky, N. 2006. User Experience – a Research Agenda. *Behaviour and Information Technology*, 25(2), pp. 91–97.

Hayes, B. 2008. Cloud computing, *Commun ACM*, vol. 51, pp. 9–11.

Hoanca, B. and Mock, H. 2006. "Secure graphical password system for high traffic public areas". In the Conference Proceedings of the 2006 symposium on Eye tracking research & applications, p. 35, Publisher ACM.

Holcombe, B. Government. 2004. "Smart Card Handbook. Smart Card Standards and Interoperability". http://www.idmanagement.gov/smart/information/ smartcard-handbook.doc. Retrieved on July 31, 2005.

Hollingsed, T. and Novick, D. G. 2007. "Usability Inspection Methods after 15 Years of Research and Practice". *ACM 25th International Conference on Design of Communication, SICDOC'07*, El Paso, TX. pp. 249–255.

Hollnagel, E., Ed. 2003. "Handbook of Cognitive Task Design", 1st Edition, CRC Press, Published June 1, 2003 Reference - 840 Pages.

Hom, J. 1998. "The Usability Methods Toolbox". http://usability.jameshom.com. Retrieved on March 14, 2005.

Hornbæk, K. and Frøkjær, E. 2004. "Two psychology-based usability inspection techniques studied in a diary experiment". *Proceedings of the Third Nordic Conference in Human Computer Interaction – NordiCHI 2004*, pp. 3–12.

Howell, W. C. and Cooke, N. J. 1989. "Training the Human Information Processor: A look at Cognitive Models". In I. Goldstein (Ed.), *Training and Development in Work Organizations: Frontier Series of Industrial and Organizational Psychology*, 3, New York: Jossey Bass. pp. 121–182.

Hutchins, E. 1995. *Cognition in the Wild*. Cambridge, MA: MIT Press.

IBM, Inc. 2006. "A usability teaching tool to demonstrate poor interface design: EasyChart". http://www-306.ibm.com/ibm/easy/eou_ext.nsf/publish/3072. Retrieved on July 8, 2006.

Idera. 2010. *Idera SharePoint. Backup and Recover SharePoint* (Version 2.7). Software. Houston, TX: Idera.

IEEE. 1998. "IEEE Standard for Software Quality Metrics Methodology". IEEE Std 1061-1998. Online Standard. http://standards.ieee.org/reading/ieee/std_public/description/se/1061-1992_desc.html. Retrieved on March 13, 2006.

IEEE. 1999. "IEEE Standard for Information technology – Telecommunications and information exchange between systems". Local and metropolitan area networks—Specific requirements – Part 11: Wireless LAN Medium Access Control (MAC) and Physical Layer (PHY) specifications – Amendment 1: High-speed Physical Layer in the 5 GHz band". IEEE Std 802.11-1999. Online Standard. http://ieeexplore.ieee.org/xpl/freeabs_all.js?arnumber=1389197. Retrieved on July 5, 2005.

International Organization for Standardization (ISO)/International Electric Committee (IEC). 1998. "Ergonomic requirements for office work with visual display terminals (VDTs) – Part 11: Guidance on Usability". ISO 9241-11. 1998. Online Standard. http://www.iso.org/iso/catalogue_detail.htm?csnumber=16883. Retrieved on November 30, 2006.

International Organization for Standardization (ISO) 9241–210:2011. Ergonomics of human system interaction - Part 210: Human-centred design for interactive systems (formerly known as 13407). International Organization for Standardization (ISO). Switzerland.

International Organization for Standardization (ISO)/International Electric Committee (IEC) 1998. ISO 9241-11. Ergonomic Requirements for Office Work with Visual Display Terminals (VDTs).

International Organization for Standardization (ISO)/ International Electric Committee (IEC) 1999. ISO/IEC 9797-1:1999 Information technology – Security techniques – Message Authentication Codes (MACs) – Part 1: Mechanisms using a block cipher.

International Organization for Standardization (ISO)/ International Electric Committee (IEC) 2001. ISO/IEC 9126. Software engineering – Product quality – Part 1: Quality model, ISO/IEC 9126-1:2001 Edition 1; 2003 Software engineering – Product quality – Part 2: External metrics, ISO/IEC TR 9126-2:2003 Edition 1; 2004 Software engineering – Product quality–Part 4: Quality in use metrics, ISO/IEC TR 9126-4:2004 Edition 1. ISO 9564–1:2002.

International Organization for Standardization (ISO)/ International Electric Committee (IEC) 2004. "ISO/IEC 13335 Information technology – Security techniques – Management of information and communications technology security."

International Organization for Standardization (ISO)/International Electric Committee (IEC) 2005. ISO/IEC 15408:2005 Information technology – Security techniques – Evaluation criteria for IT security (Common Criteria v3.0).

International Organization for Standardization (ISO)/ International Electric Committee (IEC) 25000: 2005. Software Engineering – Software Product Quality Requirements and Evaluation (SQuaRE) – Guide to SQuaRE.

International Organization for Standardization (ISO)/International Electric Committee (IEC) 2009. ISO/IEC 27004:2009 – Information technology – Security techniques – Information security management – Measurement.

ISO. 1999. "Human-Centred Design Processes for Interactive Systems". ISO 13407: 1999. Online Standard. http://www.iso.org/iso/ catalogue_detail.htm?csnumber=21197. Retrieved on May 24, 2007.

ISO. 2002. "Banking – Personal Identification Number (PIN) management and security – Part 1: Basic principles and requirements for online PIN handling in ATM and POS systems". ISO 9564-1:2002. Online Standard. http://www.iso.org/iso/catalogue_detail?csnumber=29374. Retrieved on June 4, 2007.

ISO. 2010. International Organization for Standardization (ISO)/International Electric Committee (IEC) 9241-210:2010 "Ergonomics of human-system interaction – Part 210: Human-centred design for interactive systems" <https://www.iso.org/standard/52075.html>

ISO/IEC. 2001. Software engineering – Product quality – Part 1: Quality model, ISO/IEC 9126-1:2001 Edition 1; 2003 Software engineering – Product quality – Part 2: External metrics, ISO/IEC TR 9126-2:2003 Edition 1; 2004 Software engineering -\ Product quality–Part 4: Quality in use metrics, ISO/IEC TR 9126-4:2004 Edition 1. ISO 9564-1:2002. Online Standard. http://www.iso.org/iso/catalogue_detail.htm?csnumber=22749. Retrieved on November 11, 2006.

ISO/IEC. 2011. International Organization for Standardization (ISO)/International Electrotechnical Commission (IEC) 25010:2011 "Systems and software engineering - Systems and software Quality Requirements and Evaluation (SQuaRE) - System and software quality models" <https://www.iso.org/standard/35733.html>

I/O Software Inc. 2005. "Levels of Security". http://www.iosoftware.com/pages/Support/Authentication%20Basics/Selection%20Process/index.asp. Retrieved on June 15, 2006.

Jain, A. K. 2004. "Biometric Recognition: How do I Know Who You Are?". *Proceedings of IEEE 12th Signal Processing and Communications Applications Conference*, Kusadasi, Turkey. 3540, pp. 3–5.

James, W. 1890. *Principles of Psychology*. New York, NY: Henry Holt and Company.

Janney Montgomery Scott LLC. 2005. "Janney Montgomery Scott Report". http://www. jmsonline.com/jms/ Retrieved on July 1, 2005.

Jeffries, R., Miller, J. R., Wharton, C., and Uyeda, K. M. 1991. "User Interface Evaluation in the Real World: A Comparison of Four Techniques". *Proceedings of the ACM CHI'91 Human Factors in Computing Systems Conference*, New Orleans, LA. pp. 119–124.

John, B. E. and Kieras, D. E. 1994. "The GOMS Family of Analysis Techniques: Tools for Design and Evaluation". Human-Computer Interaction Institute Technical Report CMU. ftp://www.eecs.umich.edu/people/kieras/GOMS/John-Kieras-TR94. pdf Retrieved on January 6, 2004.

John, B. E. and Kieras, D. E. 1996a. "Using GOMS for User Interface Design and Evaluation: Which Technique?". *ACM Transactions on Computer-Human Interaction*, 3(287), p. 319.

John, B. E. and Kieras, D. E. 1996b. "The GOMS Family of User Interface Analysis Techniques: Comparison and Contrast". *ACM Transactions on Computer-Human Interaction*, 3(4), pp. 320–351.

Jokela, T., Koivumaa, J., et al. (2006). "Methods for Quantitative Usability Requirements: A Case Study on the Development of the User Interface of a Mobile Phone." *Personal and Ubiquitous Computing,* 10, pp. 345–355.

Jøsang, A. and Patton, M. A. 2003. "User Interface Requirements for Authentication of Communication". *Proceedings of the Fourth Australasian user interface conference on User interfaces 2003*, Adelaide, Australia. 18, pp. 75–80.

Jøsang, A., Alfayyadh, B., Zomai, M., McNamara, J. and Grandison, T. 2007a. "Security Usability Principles for Vulnerability Analysis and Risk Assessment". *Proceedings of the Twenty-Third Annual Computer Security Applications Conference (ACSAC 2007)*, Miami Beach, Florida. pp. 269–278.

Jøsang, A., Zomai, M. and Suriadi, S. 2007b. "Usability and Privacy in Identity *Management* Architectures". Australasian Information Security Workshop: Privacy Enhancing Technologies (AISW) session, *Proceedings of the Fifth Australasian Symposium on ACSW Frontiers*, Darlinghurst, Australia. 68, pp. 143–152.

Juristo, N., Moreno, A., et al. (2007). "Analysing the Impact of Usability on Software Design." *The Journal of Systems and Software*, 80, pp. 1506–1516.

Kantner, L. and Keirnan, T. 2003. "Field Research in Commercial Product Development". *Proceedings of Usability Professionals Association (UPA) 2003: Ubiquitous Usability – Advanced Topic Seminars*, Scottsdale, AZ.

Karat, J. and Bennett, J. L. 1991. "Working Within the Design Process: Supporting Effective and Efficient Design". In Carroll, J. M. (Ed.), *Designing Interaction: Psychology at the Human-Computer Interface*, Cambridge, England: Cambridge University Press. pp. 269–285.

Karat, C., Campbell, R. and Fiegel, T. 1992. "Comparison of empirical testing and walk-through methods in user interface evaluation". *Proceedings of CHI'92 Human Factors in Computing Systems Conference*, Monterey, CA. pp. 397–404.

Kassoff, M., Kato, D., Mohsin, W. 2003. Creating GUIs for Web Services. *IEEE Internet Computing*, 7(5), pp. 66–73.

Kerckhoffs, A. 1883. "La cryptographie militaire/Military Cryptography". *Journal des sciences militaires*, 9, p. 161–191.

Kieras, D. E. 1996. "A Guide to GOMS Model Usability, Evaluation using NGOMSL". ftp://ftp.eecs.umich.edu/people/kieras/GOMS/NGOMSL_Guide.pdf Retrieved on February 3, 2009.

Kieras, D. 2001. "Using the Keystroke-Level Model to Estimate Execution Times". ftp:// ftp.eecs.umich.edu/people/kieras/GOMS/KLM.pdf Retrieved on February 3, 2009.

Kieras, D. E. 2006. "A Guide to GOMS Model Usability Evaluation using GOMSL and GLEAN4". Ann Harbor, MI: University of Michigan.

Kieras, D. E. and Bovair, S. 1984. "The Acquisition of Procedures from Text: a Production-system Analysis of Transfer of Training". *Journal of Memory and Language*, 25, pp. 507–524.

Kieras, D. E., and Polson, P. G. 1985. "An Approach to the Formal Analysis of User Complexity". *International Journal of Man-Machine Studies*, 22, pp. 365–394.

Kieras, D. E., Wood, S. D., Abotel, K. and Hornof, A. 1995. "GLEAN: A Computer-Based Tool for Rapid GOMS Model Usability Evaluation of User Interface Designs". *UIST'95 Proceedings of the ACM Symposium on User Interface Software and Technology*, Pittsburgh, PA. pp. 91–100.

Kieras, D., Wood, S. and Meyer, D. 1995. "Predictive Engineering Models Using the EPIC Architecture for a High-Performance Task". In Proceedings of CHI'95 HumanFactors in Computing Systems (Denver, May 7-11, 1995), New York: ACM.

Kieras, D., Wood, S. and Meyer, D. 1997. "Predictive Engineering Models Based on the EPIC Architecture for a Multimodal High-Performance Human-Computer Interaction Task". *ACM Transactions on Computer Human Interaction*. 4(3), pp. 230–275. New York, NY: ACM Press.

Kirakowski, J. 2001. "SUMI Questionnaire". Human Factors Research Group, Cork, Ireland: University College Cork.

Kirwan, B. and Ainsworth, L. K. 1992. "A Guide to Task Analysis" book. January 1992. doi: 10.1201/b16826, Publisher: Taylor & Francis.

Kjeldskov, J., Skov, M.B. and Stage, J. 2010. "A Longitudinal Study of Usability in Health Care: Does Time Heal?". *International Journal of Medical Informatics*, 79(6), pp. 135–143.

Kosslyn, S. M. 1980. *Image and Mind*. Cambridge, MA: Harvard University Press.

Kosslyn, S. M. 1983. *Ghosts in the Mind's Machine: Creating and Using Images in the Brain*. New York, NY: Norton.

Laird, J. E. 2008. "Frontiers in Artificial Intelligence and Applications". *Proceedings of the 2008 conference on Artificial General Intelligence*, University of Memphis, TN. 171, pp. 224–235, Amsterdam, The Netherlands: IOS Press.

Laird, J. E. and Congdon, C. B. 2009. "SOAR Users Manual". University of Michigan, MI http://soar.googlecode.com/svn/tags/Soar-Suite-9.3.0/Core/Documentation/ SoarManual.pdf Retrieved on July 22, 2009.

Laird, J. E., Newell, A. and Rosenbloom, P. S. 1987. "SOAR: An Architecture for General Intelligence". *Artificial Intelligence*, 33(1), p. 64.

Law, E. L. C., Hvannberg, E. and Cockton, G. (Eds.). 2008. *Maturing Usability: Quality in Software, Interaction and Value*. London, UK: Springer-Verlag, p. 469.

Lecerof, A. and Paternò, F. 1998. "Automatic Support for Usability Evaluation". *IEEE Transactions on Software Engineering*, 24(10), pp. 863–888.

Lepreux, S., Vanderdonckt, V., Michotte, B. 2003. Visual Design of User Interfaces by (De) composition. *Proceedings DSV-IS 2006*, LNCS, Springer, pp.157–170.

Limbourg Q., Vanderdonckt J., Michotte B., Bouillon L., Lopez-Jaquero V. 2004. USIXML: A Language Supporting Multi-path Development of User Interfaces. *EHCI/DS-VIS*, 2004, pp. 200–220.

Luyten, K., Abrams, M., Vanderdonckt, J., and Limbourg, Q. 2004. Developing User Interfaces with XML: Advances on User Interface Description Languages, Advanced Visual Interfaces 2004. Gallipoli, Italy.

Maffezzini, I. P. 2006. "Interfaces et Frontières ou Comment Passer de L'autre Côté sans se Faire Mal". Canada: Montreal, University of Quebec at Montreal.

Maltoni, D., Maio, D., Jain, A. K. and Prabhakar, S. 2003. *Handbook of Fingerprint Recognition*. New York, NY: Springer-Verlag.

Manning, B. 2009. "Highlights and Takeaways". *RSA Conference 2009 Internal Presentation*, Slide 5, San Francisco, CA.

Manolescu I., Brambilla M., Ceri S., Comai S., Fraternali P. 2005. Model-driven design and deployment of service-enabled Web applications. *ACM Transactions on Internet Technology*, 5(3), pp. 439–479.

MAP 2005. "Metodología de Análisis y Gestión de Riesgos de los Sistemas de Información (MAGERIT – v 2)."

Marr, D. and Nishihara, H. K. 1978. "Representation and Recognition of the Spatial Organization of Three-Dimensional Shapes". *Proceedings of the Royal Society of London. Series B, Biological Sciences*, 200(1140), pp. 269–294.

Martin, R.A. 2010. Common Weakness Enumeration (CWE v1.8), National Cyber Security Division of the U.S. Department of Homeland Security.

Mastin, L. 2010. Sensory Memory – Types of Memory – The Human Memory. [online] Human-memory.net. Available at: <http://www.human-memory.net/types_sensory. html> Accessed 29 Mar. 2018.

Mathy, F. and Feldman, J. 2012. What's magic about magic numbers? Chunking and data compression in short-term memory. Cognition, 122(3), 346–362.

McGraw, G. 2008. Automated Code Review Tools for Security. How Things Work.

Mell, P. and Scarfone, K., 2007. A Complete Guide to the Common Vulnerability Scoring System (CVSS 2.0), NIST and Carnegie Mellon University.

Microsoft Corporation. 2005. "Web Service Security: Scenarios, Patterns, and Implementation Guidance for Web Services Enhancements (WSE) 3.0." http://msdn. microsoft.com/en-us/library/aa480547.aspx Retrieved on September 29, 2006.

Microsoft Corporation. 2009. "Windows 2000 Server. Security with Smart Cards". http://technet.microsoft.com/en-us/library/cc962052.aspx Retrieved on July 22, 2009.

Miller, G. A. 1956. "The Magical Number Seven, Plus or Minus Two: Some Limits on our Capacity for Processing Information". *Psychological Review*, 63, p. 81–97.

Miller, M. 2008. *Cloud Computing: Web-Based Applications That Change the Way You Work and Collaborate Online*. Que Publishing.

Mitnick, K. and Simon, W. 2002. *The Art of Deception: Controlling the Human Element of Security*. Chichester, England: John Wiley & Sons.

Molich, R. and Dumas, J. S. 2008. "Comparative usability evaluation (CUE-4)". *Behaviour & Information Technology*, Vol. 27, No. 3, pp. 263–281.

Molich, R. and Nielsen, J. 1990. "Ten Usability Heuristics". http://www.useit.com/ papers/ heuristic/heuristic_list.html Retrieved on April 27, 2006.

Naur, P. 1995. *Knowing and the Mystique of Logic and Rules*. Dordrecht, Netherlands: Kluwer Academic Publishers.

Naur, P. 1998. "Human Knowing, Language, and Discrete Structures". In Naur, P. 1992, *Computing: A Human Activity*", New York, NY: ACM Press/Addison Wesley. pp. 518–535.

Naur, P. 2000. "CHI and Human Thinking". *Proceedings of the First Nordic Conference on Computer-Human Interaction – NordiCHI*, Stockholm, Sweden. pp. 23–25. Also in [50-2005] Appendix 1, pp. 199–207.

National Computer Security Center (NCSC). 1983. "Guide to Understanding Identification & Authorization in Trusted Systems". 1(5), pp. 235–479.

National Institute of Standards and Technology (NIST). 2001. "Announcing the Advance Encryption Standard (AES)". Online Standard. http://csrc.nist.gov/publications/fips/fips197/fips-197.pdf. Retrieved on June 11, 2007.

Neisser, U. 1964. "Visual Search". *Scientific American*, 210(6), pp. 94–102.

Newell, A., Rosenbloom, P. S. and Laird, J. E. 1987. "SOAR: An Architecture for General Intelligence". *Artificial Intelligence*, 33(1) pp. 1–64.

Newell, A. and Simon, H. 1997. "Computer Science and Empirical Inquiry: Symbols and Search". In John Haugeland (Ed.), *Mind Design II*, Cambridge, MA: MIT Press. pp. 81–110.

Nichols, E. A. and Peterson, G., 2007. "A Metrics Framework to Drive Application Security Improvement." *IEEE Security and Privacy.*

Nielsen, J. 1992. "Finding Usability Problems through Heuristic Evaluation". *Proceedings of ACM Computer Human Interaction (CHI'92)*, Monterey, CA. pp. 373–380.

Nielsen, J. 1993. *Usability Engineering.* Boston, MA: AP Professional.

Nielsen, J. 1994a. "Heuristic Evaluation". In Nielsen, J. and Mack, R. L. (Eds.), *Usability Inspection Methods*, New York, NY: John Wiley & Sons.

Nielsen, J. 1994b. "Heuristic Evaluation, How to Conduct a Heuristic Evaluation". http://www.useit.com/papers/heuristic/heuristic_evaluation.html Retrieved on October 9, 2003.

Nielsen, J. 2000. "Security & Human Factors". Jakob Nielsen's Alertbox. http://www.useit.com/alertbox/20001126.html Retrieved on April 10, 2004.

Nielsen, J. 2006. "Severity Ratings for Usability Problems". http://www.useit.com/ papers/heuristic/severityrating.html Retrieved on January 13, 2007.

National Institute of Standards and Technology (NIST). 2001. "Announcing the Advance Encryption Standard (AES)". Online Standard. http://csrc.nist.gov/publications/fips/fips197/fips-197.pdf. Retrieved on June 11, 2007.

Norman, D. A. 1983. "Some Observations on Mental Models". In Gentner, D. and Stevens, A. L. (Eds.), *Mental Models.* Hillsdale, NJ: L. Erlbaum Associates, pp. 7–14.

Norman, Donald A. 1988. *The Design of Everyday Things.* New York, NY: Doubleday.

Nuxoll, A. and Laird, J. 2004. "A Cognitive Model of Episodic Memory Integrated With a General Cognitive Architecture". *International Conference on Cognitive Modeling.* Pittsburgh, PA: pp. 22–225.

OECD 2005. *"The Promotion of a Culture of Security for Information Systems and Networks in OECD Countries.* DSTI/ICCP/REG(2005)1/FINAL, Organisation for Economic Co-operation and Development.".

Olson, J. R. and Olson, G. M. 1990. "The Growth of Cognitive Modeling in Human-Computer Interaction since GOMS". *Human-Computer Interaction*, 5, pp. 221–265.

Open Web Application Security Project (OWASP). 2009. "SQL Injection Vulnerabilities – SQL Injection Prevention Cheat Sheet". http://www.owasp.org/index.php/SQL_Injection_Prevention_Cheat_Sheet Retrieved on July 28, 2009.

O'Regan, J. K. and Noë, A. 2001. "A Sensorimotor Account of Vision and Visual Consciousness". *Behavioural and Brain Sciences*, 24, pp. 939–1031, Cambridge, MA: Cambridge University Press.

Passfaces Corporation. 2009. "Passface® Authentication Method". http://www.passfaces.com Retrieved on January 9, 2009.

Paternò, F. and Santoro, C. 2008. "Remote Usability Evaluation: Discussion of a General Framework and Experiences". In Law, E. L.-C., Hvannberg, E. and Cockton, G. (Eds.). *Maturing Usability: Quality in Software, Interaction and Value.* London, UK: Springer-Verlag, 469.

Penn, J. 2008. "The State of Enterprise IT Security 2008 to 2009". Business Data Services North America and Europe. Forrester Research survey. http://www.forrester.com/rb/Research/state_of_enterprise_it_security_2008_to/q/id/47857/t/2 Retrieved on February 2, 2009.

Perks, M. 2003. "Best practices for software development projects". developerWorks, Technical Library. http://www.ibm.com/developerworks/websphere/library/techarticles/0306_perks/perks2.html

Perlman, G. 2006. "Web-Based User Interface Evaluation with Questionnaires". http://www.acm.org/~Eperlman/question.html Retrieved on June 12, 2006.

Pfleeger, S. L. 2009. *Useful Cybersecurity Metrics.* computer.org/ITPro, IEEE C omputer S ociet y: 38–45.

Pierotti, D. 1996. "Usability Techniques – Heuristic Evaluation: A System Checklist", Xerox Corporation, Version 1.0. http://www.stcsig.org/usability/ topics/articles/he-checklist.html Retrieved on March 12, 2004.

Pinna-Dery, A. M., Fierstone, J., Picard, E. 2003. Component Model and Programming: A First Step to Manage Human Computer Interaction Adaptation. *Mobile HCI.* 2003, pp. 456–465.

Poirier, P., Hardy-Vallée, B. and DePasquale, J. F. 2005. "Embodied Categorization". In Henri, C. and Claire, L. (Eds.), *Handbook of Categorization in Cognitive Science.* pp. 739–765. Oxford: Elsevier Science.

Posthumus, S., Solms, R.V. 2004. A Framework for the Governance of Information Security, *Computers and Security (COMPSEC),* 23(8):638–646.

Proctor, R. W. and Vu, K.-P. L. 2005. *Handbook of Human Factors in Web Design.* Mahwah, NJ: L. Erlbaum Associates. p. 396.

Rad, P.M., Badashian, A.S., Meydanipour, G., Delcheh, M.A., Alipour M., Afzali, H. 2004. A Survey of Cloud Platforms and their Future, *Lecture Notes in Computer Science,* 5592, pp. 788–796.

Radha, N. and Kavitha, A. 2012. "Rank level fusion using fingerprint and iris biometrics," *Indian Journal of Computer Science and Engineering,* vol. no. 6, pp. 917–923, 2012.

Renaud, K. 2003. "Quantifying the Quality of Web Authentication Mechanisms – A Usability Perspective". *Journal of Web Engineering,* 3(2), pp. 95–123.

Raphael, B., Bhatnagar, G., and Smith, I. F. 2002. Creation of Flexible Graphical user Interfaces through Model Composition. *Artif. Intell. Eng. Des. Anal. Manuf.* 16, 3, pp. 173–184. doi: http://dx.doi.org/10.1017/S0890060402163049.

Richardson, D.J. 2000. "Cognitive Walkthroughs, Human-Computer Interaction". I Oakland, CA: CS 121 Topic 6: Cognitive Walkthroughs. University of California,.

Ritter, F. E. 2004. "Choosing and Getting Started with a Cognitive Architecture to Test and Use Human-Machine Interfaces". *MMI-Interaktiv Journal,* 7(17), p. 37.

RSA Security LLC. 2010a. Bedford, MA.

RSA Security LLC. 2010b. "Enterprise Single Sign-On. RSA Laboratories". http://www.rsa.com/rsalabs/node.asp?id=2541 Retrieved on May, 2010.

Ruitenbeek, E. V. and Scarfone, K. 2009. The Common Misuse Scoring System (CMSS): Metrics for Software Feature Misuse Vulnerabilities. NIST Interagency Report 7517, National Institute of Standards and Technology.

Saltzer, J. H. and Schroeder, M. D. 1975. "The Protection of Information in Computer Systems". *Proceedings of the IEEE*, 63(9), pp. 1278–1308.

Santoro, T. and Kieras, D. 2001. GOMS models for team performance. In J.Pharmer and J. Freeman 32 (Organizers), Complementary methods of modeling team performance. Panel presented at The 45th Annual Meeting of the Human Factors and Ergonomics Society, Minneapolis/St. Paul.

Sasse, M. A. 2004. "Usability and Security – Why We Need to Look at the Big Picture". ISS 2004, London, UK: University College London.

Sasse, M. A., Brostoff, S. and Weirich, D. 2001. "Transforming the "Weakest Link" – A Human/Computer Interaction Approach to Usable and Effective Security". *BT Technology Journal*, 19(3), pp. 122–131.

Sasse, M. A. and Flechais, I. 2005. "Usable Security – Why Do We Need It? How Do We Get It?". In Cranor, L. F. and Garfinkel, S. (Eds.). *Security and Usability: Designing Secure Systems That People Can Use*. pp. 13–30. Sebastopol, CA: O'Reilly.

Sauro, J. 2006. "Sample Size for Discovering Problems in a UI, Measuring Usability Website". http://www.measuringusability.com/samplesize/ problem_discovery.php Retrieved on April 5, 2006.

Sauro, J. and Kindlund, E. 2005 "A Method to Standardize Usability Metrics into a Single Score". http://www.measuringusability.com/SUM/p482-sauro.pdf Retrieved on May 8, 2006.

Scambray, J., Shema, M. and Sima, C. 2006. *Hacking Exposed: Web Applications*. San Francisco, CA: McGraw-Hill.

Scarfone, K., and Souppaya, M. 2009. NIST Special Publication 800-118 (Draft): Guide to Enterprise Password Management (Draft), Recommendations of the National Institute of Standards and Technology, Computer Security Division Information Technology Laboratory, National Institute of Standards and Technology. Gaithersburg, MD. April 2009.

Scholtz, J. 2003. "Usability Evaluation". National Institute of Standards and Technology. p. 8.

Schneiderman, B. 1998. *Designing the User Interface: Strategies for Effective Human-Computer Interaction*. 3rd Edition. Reading, MA: Addison-Wesley.

Schrepp, M. 2010. "GOMS analysis as a tool to investigate the usability of web units for disabled users". In Universal Access in the Information Society 9(1):77–86. March 2010.

Schrepp, M. and Fischer, P. 2006. "A GOMS Model for Keyboard Navigation in Web Pages and Web Applications". *ICCHP Conference 2006*, Linz, Austria. http://www.axistive.com/a-goms-model-for-keyboard-navigation-in-web-pages-and-web-applications.html Retrieved on November 10, 2007.

Secure Computing: 2018. "SafeNet OTP 110® PremierAccess, Strong Authentication." https://safenet.gemalto.com/multi-factor-authentication/authenticators/safenet-otp-110/ Retrieved on October 15, 2006.

Seffah, A., Abran, A., Suryn, W., Khelifi, A., Rilling, J. and Robert, F. 2003. "Consolidating the ISO Usability Models." Concordia University, Montreal, Canada.

Seffah, A. and Donyaee, M. 1998. "Metrics and Measurement of Usability." *International Encyclopedia of Ergonomics and Human Computer Interaction*. 2nd Edition, vol. 1. Boca Raton, FL: CRC Press.

Seffah A., Donyaee M., Kline R. and Padda H. K. 2006. "Usability Metrics: A Roadmap for a Consolidated Model." *Journal of Software Quality*, 14(2).

Seffah A. and Metzker, E. 2004. "The Obstacles and Myths of Usability and Software engineering: Avoiding the Usability Pitfalls Involved in Managing the Software Development Life Cycle." *Communications of the ACM*, 47(12), pp. 71–76.

Seffah, A., Kececi, N., et al. 2001. QUIM: A Framework for Quantifying Usability Metrics in Software Quality Models. *Proceedings of the Second Asia-Pacific Conference on Quality Software (APAQS'01)*.

Shadow. 2008. "Authentication Risk Analysis Worksheet Answers." http://www.scribd.com/riskAnalysis.html. Retrieved on January 2005.

Shneiderman, B. and Plaisant, B. 2005. *Designing the User Interface: Strategies for Effective Human Computer Interaction*. 4th Edition. Boston, MA: Pearson Addison-Wesley.

Smith, R. E. 2002. *Authentication: From Passwords to Public Keys*. Boston, MA: Addison-Wesley.

Smith, S. L. and Mosier, J. N. 1986. "Guidelines for Designing User Interface Software". ESD-TR, 86(27), p. 34.

Sperling, G. 1960. "The Information Available in Brief Visual Presentations". *Psychological Monographs*, 74(498), pp. 1–29.

Stajano, F. 2003. "Security For Whom? The Shifting Security Assumptions of Pervasive Computing". In M. Okada et al. (Eds.): *Software Security—Theories and Systems*, 2609, Berlin, Germany: Springer-Verlag.

Stiegler, M., Karp, A. H., Yee, K. P. and Mark Miller. 2004. "Polaris: Virus Safe Computing for Windows XP". http://www.hpl.hp.com/techreports/2004/HPL-2004-221.pdf Retrieved on May 24, 2006.

Stuerzlinger, W., Chapuis, O., Phillips, D. and Roussel, N. 2006. "User Interface Façades: Towards Fully Adaptable User Interfaces". *Proceedings of the 19th Annual ACM Symposium on User Interface Software and Technology-UIST '06*, Montreux, Switzerland. pp. 309–318. New York, NY: ACM Press.

Stufflebeam, D. L. 2000. "Guidelines for Developing Evaluation Checklists: The Checklists Development Checklist (CDC)." http://www.wmich.edu/evalctr/checklists/guidelines_cdc.pdf Retrieved on November 10, 2006.

Swanson, M., Bartol, N., Sabato, J., and Graffo, L. 2003. Security Metrics Guide for Information Technology Systems. NIST Special Publication 800-55 Revision 1, National Institute of Standards and Technologies.

Sy, D. 2009. "Usability Over Time: Longitudinal Research Studies". In *Designing the User Experience at Autodesk – Insights on Innovation, Inspiration, and the Practice of Design*. http://dux.typepad.com/dux/2009/05/usability-over-time-longitudinal-research-studies.html. Retrieved on July 22, 2010.

The MIT Kerberos Team. 2006. "Kerberos: The Network, Authentication Protocol". http://Web.Mit.Edu/Kerberos Retrieved on March 4, 2007.

Theofanos M. 2006. "A practical guide to the CIF: Usability measurements," *Interactions*, 13, pp. 34–37.

Tulving, E. and Schacter, D. L. 1990. "Priming and Human Memory Systems." *Science*, 247(4940), pp. 301–306.

Uldall-Espersen, T. 2008. The Usability Perspective Framework. CHI 2008.

United Airlines. 2007. "Delayed and damaged baggage". Baggage Services/Lost and Found. <http://www.united.com/page/article/0,6722,1037,00.html> Retrieved on March 4, 2008.

Usability Professionals Association (UPA). 2010. "Usability – Body of Knowledge, Methods, Heuristic Evaluation (How To), The Usability Body of Knowledge". http://www.usabilitybok.org/methods/p275?section=how-to Retrieved on May 5, 2010.

Väänänen, K., Väätäjä, H. and Vainio, T. 2008. Opportunities and Challenges of Designing the Service User eXperience (SUX) in Web 2.0. In: Saariluoma, P. and Isomäki, H. (Eds.). *Future Interaction Design II*. Springer.

Valentine, T: 1998. "An Evaluation of the Passfaces™ Personal Authentication System". Technical Report, Goldmsiths College, London, UK: University of London.

Van Someren, M. W., Barnard, Y. F. and Sandberg, J. A. C. 1994. *The Think Aloud Method: A Practical Guide to Modeling Cognitive Processes*. London, UK: Academic Press.

Vance, A. 2010. "If Your Password Is 123456, Just Make It HackMe", New York Times, Technology <https://www.nytimes.com/2010/01/21/technology/21password.html>

Vasco Data Security International, Inc. 2010. http://www.vasco.com/ Retrieved on May 12, 2009.

VeriSign, Inc. 2010. "VeriSign® Unified Authentication". http://www.verisign.com/static/DEV016111.pdf Retrieved on May 11, 2010.

Vermeulen J., Vandriessche Y., Clerckx T., Luyten K. and Coninx K. 2007. Service-interaction Descriptions: Augmenting Services with User Interface Models, *Proceedings Engineering Interactive Systems* 2007, Salamanca, Springer Verlag.

Voyles, R. M., Jr., and Khosla, P. K. 1995. "Tactile Gestures for Human/Robot Interaction". *Intelligent Robots and Systems 95 – Human Robot Interaction and Cooperative Robotsapos. Proceedings. 1995 IEEE/RSJ International Conference*, Pittsburgh, PA. 3(7), p. 13.

Walden, J., and Doyle, M. 2009. "Security of Open SourceWeb Applications." *Third International Symposium on Empirical Software Engineering and Measurement*, pp. 545–553.

Wang, J. A. and Wang, H. 2009. Security Metrics for Software Systems. *Proceedings of the 47th Annual Southeast Regional Conference (ACMSE '09)*.

Welie, M. V., 2008, "Patterns in Interaction Design", [Online]. Available from http://www.welie.com.

Whiteside, J., Bennett, J. and Holtzblatt, K. 1988. *Usability Engineering: Our Experience and Evolution. Handbook of Human Computer Interaction*. New York, NY: North Holland, pp. 791–817.

Whitten, A., and J. Tygar, J. D. 1998. "Usability of Security: A Case Study". CMU-CS-98-155, Pittsburgh, PA: Carnegie Mellon University.

Whitten, A. and J. Tygar. 1999. "Why Johnny Can't Encrypt: A Usability Evaluation of PGP 5.0." *Proceedings of the 8th USENIX Security Symposium*, Washington, D.C. pp. 169–184.

Wickens, C. D. 1992. *Engineering Psychology and Human Performance*. 2nd Edition Chapter 5, pp. 167–210. New York, NY: Harper Collins.

Williams, K. E. and Voigt, J. R. 2004. "Evaluation of a Computerized Aid for Creating Human Behavioral Representations of Human-Computer Interaction". *Human Factors*, 46(2), pp. 288–303.

Woodward, J. D., Orlans, N. M. and Higgins, P. T. 2003. *Biometrics – Identity Assurance in the Information Age*. Berkeley, CA: McGraw-Hill/Osborne Media. p. 432.

Wood, S. 1993. Issues in the implementation of aGOMS-model design tool. Unpublished report, Universityof Michigan.

Wood, S. D. 2000. Extending GOMS to human error and applying it to error-tolerant design. Doctoral dissertation, University of Michigan.

Wu, M., Miller, R. C. and Garfinkel, S. 2006. "Do Security Toolbars Actually Prevent Phishing Attacks?" Massachusetts Institute of Technology, Cambridge, MA.

Yee, K. P. 2004. "Aligning Security and Usability." *IEEE Security and Privacy*, 2(5), pp. 48–55.

Yee, K. P. 2005. "Guidelines and Strategies for Secure Interaction Design." In Cranor, L. F. and Garfinkel, S. (Eds.). *Security and Usability: Designing Secure Systems That People Can Use*. Sebastopol, CA: O'Reilly. p. 253–280.

Zeman, J., Tanuska, P., et al. 2009. The Utilization of Metrics Usability to Evaluate the Software Quality. *International Conference on Computer Technology and Development*.

Zhang, D. D. 2004. *Palmprint Authentication*. Norwell, MA: Kluwer Academic Publishers. p. 204.

Zou, Y., Zhang, Q., et al. 2007. "Improving the Usability of e-Commerce Applications Using Business Processes." *IEEE Transactions on Software Engineering*, 33(12), pp. 837–855.

Zurko, M. and Simon, R. 1996. "User-centered Security." *Proceedings of the UCLA Conference on New Security Paradigms Workshops*, Lake Arrowhead, CA. pp. 27–33.

Index

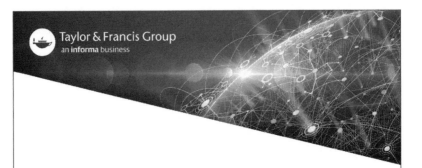

Printed and bound by CPI Group (UK) Ltd, Croydon, CR0 4YY

24/10/2024

01778284-0008